PHYSICAL GEOGRAPHY

PHYSICAL GEOGRAPHYC

by
Kabir Bassi | Armaan Andra | Onkar Bhargava

Wordpen Academics
Publishing Science That Matter

ISBN: 978-93-6084-756-2 (Hardback)

Published by : **WORDPEN ACADEMICS**
4736/23, Ansari Road, Darya Ganj,
New Delhi -110 002
Phone: +91-011-35004336
E-mail: info@wordpenacademics.com

Printed at **:** **Replika Press Pvt. Ltd.**

Exclusive Distributor: ***Bio-Green Books, New Delhi***

Preface

"Physical Geography" is a comprehensive exploration of the Earth's natural processes, landforms, climates, and environments, offering readers a holistic understanding of the physical systems that shape our planet. This book serves as an essential resource for students, educators, and enthusiasts seeking to delve into the intricacies of Earth's physical geography and the dynamic interactions that drive its ever-changing landscapes.

Physical geography is a multifaceted discipline that examines the processes and patterns of the Earth's surface, atmosphere, hydrosphere, and biosphere. From the formation of mountains and rivers to the dynamics of weather and climate, physical geography encompasses a wide range of phenomena and processes that influence the distribution of life and resources on our planet.

As editors of this volume, we have endeavored to compile a diverse array of perspectives, methodologies, and case studies from leading experts and researchers in the field of physical geography. Through a blend of descriptive text, vivid illustrations, and interactive learning tools, this book covers a wide range of topics, including geomorphology, climatology, biogeography, hydrology, and environmental science.

Our overarching goal in presenting this book is to foster a deeper appreciation and understanding of the dynamic processes and patterns that shape the Earth's physical environment. By providing readers with a comprehensive overview of physical geography concepts, theories, and applications, we aim to inspire curiosity, critical thinking, and stewardship of the natural world.

We extend our sincere gratitude to all the contributors who have generously shared their expertise, experiences, and insights in the field of physical geography. It is our hope that this book will serve as a valuable resource for students, educators, and enthusiasts alike, empowering them to explore and appreciate the wonders of the Earth's physical landscape and the dynamic processes that shape it.

Kabir Bassi | Armaan Andra | Onkar Bhargava

Contents

1

Physical Geography: An Introduction

Geography is a field of study concerned with space and spatial relations. Usually, the focus of geography concerns places, landmarks, physical landforms, climate, humans, and ecological distributions and locations on Earth. However, a small field of geography is concerned with mapping the physical geography of other bodies in our solar system.

Physical geography is a subfield of geography concerned with the physical landforms and natural environments of the Earth, the processes that form or transform them, and the spatial distribution and relationships between them. The field does not focus on human-dependant geographical concepts like national or cultural boundaries, human trade, migration routes, or human-made landmarks. Humans still play a role in physical geography when the field investigates the human impacts on the environment's natural features.

Physical Geography: Definition

Physical geography consists of many diverse elements. These include: the study of the earth's interaction with the sun, seasons, the composition of the atmosphere, atmospheric pressure and wind, storms and climatic disturbances, climate zones, microclimates, the hydrologic cycle, soils, rivers and streams, flora and fauna, weathering, erosion, natural hazards, deserts, glaciers and ice sheets, coastal terrain, ecosystems, geologic systems, and so much more.

The Four Spheres

It's a little deceiving (even overly simplistic) to say that physical geography studies the Earth as our home and looks at the four spheres because each possible area of research encompasses so much.

The atmosphere itself has several layers to study, but the atmosphere as a topic under the lens of physical geography also includes research areas such as the ozone layer, the greenhouse effect, wind, jet streams, and weather.

The hydrosphere encompasses everything having to do with water, from the water cycle to acid rain, groundwater, runoff, currents, tides, and oceans.

The biosphere concerns living things on the planet and why they live where they do, with topics from ecosystems and biomes to food webs and the carbon and nitrogen cycles.

The study of the lithosphere includes geological processes, such as the formation of rocks, plate tectonics, earthquakes, volcanoes, soil, glaciers, and erosion.

Historical evolution of the discipline

From the birth of geography as a science during the Greek classical period and until the late nineteenth century with the birth of anthropogeography (human geography), geography was almost exclusively a natural science: the study of location and descriptive gazetteer of all places of the known world. Several works among the best known during this long period could be cited as an example, from Strabo (Geography), Eratosthenes (Geographika) or Dionysius Periegetes (Periegesis Oiceumene) in the Ancient Age. In more modern times, these works include the Alexander von Humboldt (Kosmos) in the nineteenth century, in which geography is regarded as a physical and natural science through the work Summa de Geografía of Martín Fernández de Enciso from the early sixteenth century, which indicated for the first time the New World.

During the eighteenth and nineteenth centuries, a controversy exported from geology, between supporters of James Hutton (uniformitarianism thesis) and Georges Cuvier (catastrophism) strongly influenced the field of geography, because geography at this time was a natural science.

Two historical events during the nineteenth century had a great effect on the further development of physical geography. The first was the European colonial expansion in Asia, Africa, Australia and even America in search of raw materials required by industries during the Industrial Revolution. This fostered the creation of geography departments in the universities of the colonial powers and the birth and development of national geographical societies, thus giving rise to the process identified by Horacio Capel as the institutionalization of geography.

The exploration of Siberia is an example. In the mid-eighteenth century, many geographers were sent to perform geographical surveys in the area of Arctic Siberia. Among these is who is considered the patriarch of Russian geography, Mikhail Lomonosov. In the mid-1750s Lomonosov began working in the Department of Geography, Academy of Sciences to conduct research in Siberia. They showed the organic origin of soil and developed a comprehensive law on the movement of the ice, thereby founding a new branch of geography: glaciology. In 1755 on his initiative was founded Moscow University where he promoted the study of geography and the training of geographers. In 1758 he was appointed director of the Department of Geography, Academy of Sciences, a post from which would develop a working methodology for geographical survey guided by the most important long expeditions and geographical studies in Russia.

The contributions of the Russian school became more frequent through his disciples, and in the nineteenth century we have great geographers such as Vasily

Dokuchaev who performed works of great importance as a "principle of comprehensive analysis of the territory" and "Russian Chernozem". In the latter, he introduced the geographical concept of soil, as distinct from a simple geological stratum, and thus found a new geographic area of study: pedology. Climatology also received a strong boost from the Russian school by Wladimir Köppen whose main contribution, climate classification, is still valid today. However, this great geographer also contributed to the paleogeography through his work "The climates of the geological past" which is considered the father of paleoclimatology. Russian geographers who made great contributions to the discipline in this period were: NM Sibirtsev, Pyotr Semyonov, K.D. Glinka, Neustrayev, among others.

The second important process is the theory of evolution by Darwin in mid-century (which decisively influenced the work of Friedrich Ratzel, who had academic training as a zoologist and was a follower of Darwin's ideas) which meant an important impetus in the development of Biogeography.

Another major event in the late nineteenth and early twentieth centuries took place in the United States. William Morris Davis not only made important contributions to the establishment of discipline in his country but revolutionized the field to develop cycle of erosion theory which he proposed as a paradigm for geography in general, although in actually served as a paradigm for physical geography. His theory explained that mountains and other landforms are shaped by factors that are manifested cyclically. He explained that the cycle begins with the lifting of the relief by geological processes (faults, volcanism, tectonic upheaval, etc.). Factors such as rivers and runoff begin to create V-shaped valleys between the mountains (the stage called "youth"). During this first stage, the terrain is steeper and more irregular. Over time, the currents can carve wider valleys ("maturity") and then start to wind, towering hills only ("senescence"). Finally, everything comes to what is a plain flat plain at the lowest elevation possible (called "baseline") This plain was called by Davis' "peneplain" meaning "almost plain" Then river rejuvenation occurs and there is another mountain lift and the cycle continues.

Although Davis's theory is not entirely accurate, it was absolutely revolutionary and unique in its time and helped to modernize and create a geography subfield of geomorphology. Its implications prompted a myriad of research in various branches of physical geography. In the case of the Paleogeography, this theory provided a model for understanding the evolution of the landscape. For hydrology, glaciology, and climatology as a boost investigated as studying geographic factors shape the landscape and affect the cycle. The bulk of the work of William Morris Davis led to the development of a new branch of physical geography: Geomorphology whose contents until then did not differ from the rest of geography. Shortly after this branch would present a major development. Some of his disciples made significant contributions to various branches of physical geography such as Curtis Marbut and his invaluable legacy for Pedology, Mark Jefferson, Isaiah Bowman, among others.

Sub-branches

Physical geography can be divided into several branches or related fields, as follows:

Geomorphology

It is concerned with understanding the surface of the Earth and the processes by which it is shaped, both at the present as well as in the past. Geomorphology as a field has several sub-fields that deal with the specific landforms of various environments e.g. desert geomorphology and fluvial geomorphology; however, these sub-fields are united by the core processes which cause them, mainly tectonic or climatic processes. Geomorphology seeks to understand landform history and dynamics, and predict future changes through a combination of field observation, physical experiment, and numerical modeling (Geomorphometry). Early studies in geomorphology are the foundation for pedology, one of two main branches of soil science.

Hydrology

It is predominantly concerned with the amounts and quality of water moving and accumulating on the land surface and in the soils and rocks near the surface and is typified by the hydrological cycle. Thus the field encompasses water in rivers, lakes, aquifers and to an extent glaciers, in which the field examines the process and dynamics involved in these bodies of water. Hydrology has historically had an important connection with engineering and has thus developed a largely quantitative method in its research; however, it does have an earth science side that embraces the systems approach. Similar to most fields of physical geography it has sub-fields that examine the specific bodies of water or their interaction with other spheres e.g. limnology and ecohydrology.

Glaciology is the study of glaciers and ice sheets, or more commonly the cryosphere or ice and phenomena that involve ice. Glaciology groups the latter (ice sheets) as continental glaciers and the former (glaciers) as alpine glaciers. Although research in the areas is similar to research undertaken into both the dynamics of ice sheets and glaciers, the former tends to be concerned with the interaction of ice sheets with the present climate and the latter with the impact of glaciers on the landscape. Glaciology also has a vast array of sub-fields examining the factors and processes involved in ice sheets and glaciers e.g. snow hydrology and glacial geology.

Biogeography

It is the science which deals with geographic patterns of species distribution and the processes that result in these patterns. Biogeography emerged as a field of study as a result of the work of Alfred Russel Wallace, although the field prior to the late twentieth century had largely been viewed as historic in its outlook and descriptive in its approach. The main stimulus for the field since its founding has been that of evolution, plate tectonics and the theory of island biogeography. The field can largely be divided into five sub-fields: island biogeography, paleobiogeography, phylogeography, zoogeography and phytogeography.

Climatology

It is the study of the climate, scientifically defined as weather conditions averaged over a long period of time. Climatology examines both the nature of micro (local) and (global) climates and the natural and anthropogenic influences on them. The also sub-divided largely into the climates of various regions and the study of c phenomena or time periods e.g. tropical cyclone rainfall climatology and imatology.

Soil geography deals with the distribution of soils across the terrain. discipline is fundamental to both physical geography and pedology.

Pedology is the study of soils in their natural environment. It deals with pedogenesis, soil morphology, soil classification. Soil geography studies the spatial distribution of soils as it relates to topography, climate (water, air, temperature), soil life (micro-organisms, plants, animals) and mineral materials within soils (biogeochemical cycles).

Palaeogeography

It is a cross-disciplinary study that examines the preserved material in the stratigraphic record to determine the distribution of the continents through geologic time. Almost all the evidence for the positions of the continents comes from geology in the form of fossils or paleomagnetism. The use of these data has resulted in evidence for continental drift, plate tectonics, and supercontinents. This, in turn, has supported palaeogeographic theories such as the Wilson cycle.

Coastal geography is the study of the dynamic interface between the ocean and the land, incorporating both the physical geography (i.e. coastal geomorphology, geology, and oceanography) and the human geography of the coast. It involves an understanding of coastal weathering processes, particularly wave action, sediment movement and weathering, and also the ways in which humans interact with the coast. Coastal geography, although predominantly geomorphological in its research, is not just concerned with coastal landforms, but also the causes and influences of sea level change.

Oceanography

is the branch of physical geography that studies the Earth's oceans and seas. It covers a wide range of topics, including marine organisms and ecosystem dynamics (biological oceanography); ocean currents, waves, and geophysical fluid dynamics (physical oceanography); plate tectonics and the geology of the sea floor (geological oceanography); and fluxes of various chemical substances and physical properties within the ocean and across its boundaries (chemical oceanography). These diverse topics reflect multiple disciplines that oceanographers blend to further knowledge of the world ocean and understanding of processes within it.

Quaternary science

is an interdisciplinary field of study focusing on the Quaternary period, which encompasses the last 2.6 million years. The field studies the last ice age and the recent interstadial the Holocene and uses proxy evidence to reconstruct the past environments during this period to infer the climatic and environmental changes that have occurred.

Landscape ecology is a sub-discipline of ecology and geography that address how spatial variation in the landscape affects ecological processes such as the distribution and flow of energy, materials, and individuals in the environment (w in turn, may influence the distribution of landscape "elements" themselves su hedgerows). The field was largely funded by the German geographer Carl T Landscape ecology typically deals with problems in an applied and holistic cont The main difference between biogeography and landscape ecology is that the latter is concerned with how flows or energy and material are changed and their impacts on the landscape whereas the former is concerned with the spatial patterns of species and chemical cycles.

Geomatics is the field of gathering, storing, processing, and delivering geographic information, or spatially referenced information. Geomatics includes geodesy (scientific discipline that deals with the measurement and representation of the earth, its gravitational field, and other geodynamic phenomena, such as crustal motion, oceanic tides, and polar motion), geographical information science (GIS) and remote sensing (the short or large-scale acquisition of information of an object or phenomenon, by the use of either recording or real-time sensing devices that are not in physical or intimate contact with the object).

Environmental geography is a branch of geography that analyzes the spatial aspects of interactions between humans and the natural world. The branch bridges the divide between human and physical geography and thus requires an understanding of the dynamics of geology, meteorology, hydrology, biogeography, and geomorphology, as well as the ways in which human societies conceptualize the environment. Although the branch was previously more visible in research than at present with theories such as environmental determinism linking society with the environment. It has largely become the domain of the study of environmental management or anthropogenic influences.

Difference between Physical Geography and Human Geography

Physical geography is that branch of natural science which deals with the study of processes and patterns in the natural environment like the atmosphere,biosphere and geosphere, as opposed to the cultural or built environment, the domain of human geography.

Within the body of physical geography, the Earth is often split either into several spheres or environments, the main spheres being the atmosphere, biosphere, cryosphere, geosphere, hydrosphere, lithosphere and pedosphere. Research in physical geography is often interdisciplinary and uses the systems approach.

Human geography is manufactured items and how things are produced using humans help and physical geography is the creation of environment and natural production.

Human geography is the study of how humans have affected and been affected by the environment and why civilizations are where they are. It talks about culture, arithmetic density, migration, population, etc. Physical geography, however, is the study of the physical characteristics of Earth (minus the human aspect), such as mountains, oceans, rivers, etc.

Why Study Physical Geography:

1. To understand basic physical systems that affect everyday life (e.g. earth-sun relationships, water cycle, wind and ocean currents).
2. To learn the location of places and the physical and cultural characteristics of those places in order to function more effectively in our increasingly interdependent world.
3. To develop a mental map of your community, state, country and the world so that you can understand the "where" of places and events.
4. To understand the geography of past times and how geography has played important roles in the evolution of people, their ideas, places and environments.
5. To explain how the processes of human and physical systems have arranged and sometimes changed the surface of the Earth.
6. To understand the spatial organization of society and see order in what often appears to be random scattering of people and places.
7. To recognize spatial distributions at all scales – local to worldwide – in order to understand the complex connectivity of people and places.
8. To be able to make sensible judgments about matters involving relationships between the physical environment and society.
9. To appreciate Earth as the home of humankind and provide insight for wise management decisions about how the planet's resources should be used.
10. To understand global interdependence and to become a better global citizen.

Four Spheres of the Earth

Earth is blanketed by an atmosphere consisting of 78.0% nitrogen, 20.9% oxygen, and 0.92% Argon. The atmosphere has five layers: troposphere, stratosphere, mesosphere, thermosphere, and exosphere. 75% of the atmosphere's gases are in the bottom-most layer, the troposphere.

The magnetic field created by the internal motions of the core produces the magnetosphere which protects the Earth's atmosphere from the solar wind. As the earth is 4.5 billion years old, it would have lost its atmosphere by now if there were no protective magnetosphere.

The atmosphere is composed of 78% nitrogen and 21% oxygen. The remaining one percent contains small amounts of other gases including CO2 and water vapors. Water vapors and CO2 allow the Earth's atmosphere to catch and hold the Sun's energy through a phenomenon called the greenhouse effect. This allows Earth's surface to be warm enough to have liquid water and support life.

In addition to storing heat, the atmosphere also protects living organisms by shielding the Earth's surface from cosmic rays. Note that the level of protection is high enough to prevent cosmic rays from destroying all life on Earth, yet low enough to aid the mutations that have an important role in pushing forward diversity in the biosphere

The area near the surface of the earth can be divided up into four inter-connected "geo-spheres:" the lithosphere, hydrosphere, biosphere, and atmosphere. Scientists can classify life and material on or near the surface of the earth to be in any of these four spheres.

The names of the four spheres are derived from the Greek words for stone (litho), air (atmo), water (hydro), and life (bio).

Lithosphere

The lithosphere is the solid, rocky crust covering entire planet. This crust is inorganic and is composed of minerals. It covers the entire surface of the earth from the top of Mount Everest to the bottom of the Mariana Trench.

Hydrosphere

The hydrosphere is composed of all of the water on or near the earth. This includes the oceans, rivers, lakes, and even the moisture in the air. Ninety-seven percent of the earth's water is in the oceans. The remaining three percent is fresh water; three-quarters of the fresh water is solid and exists in ice sheets

Biosphere

The biosphere is composed of all living organisms. Plants, animals, and one-celled organisms are all part of the biosphere. Most of the planet's life is found from three meters below the ground to thirty meters above it and in the top 200 meters of the oceans and seas.

Atmosphere

The atmosphere is the body of air which surrounds our planet. Most of our atmosphere is located close to the earth's surface where it is most dense. The air of our planet is 79% nitrogen and just under 21% oxygen; the small amount remaining is composed of carbon dioxide and other gasses.

All four spheres can be and often are present in a single location. For example, a piece of soil will of course have mineral material from the lithosphere. Additionally, there will be elements of the hydrosphere present as moisture within the soil, the biosphere as insects and plants, and even the atmosphere as pockets of air between soil pieces.

Indian Physical Geography

India occupies the greater part of a peninsular subcontinent fronting on two great arms of the Indian Ocean—the Arabian Sea on the west and the Bay of Bengal on the east. Just off its southern coast lies the island nation of Sri Lanka, separated from the mainland by the Palk Strait. Land boundaries are shared with Pakistan, China. Nepal, Bhutan, Burma, and Bangladesh. On the northwestern frontier is the disputed territory of Kashmir; although India claims all of Kashmir, it holds only the southern part, the rest being occupied by Pakistan.

Land

India has three distinct physical regions: the Himalayas, the Ganges Plain, and the Deccan Plateau. The Himalayas extend along the nation's northern border, forming a high mountain wall, 100 to 150 miles (160 to 240 km) wide, that separates the Indian subcontinent from Asia's interior. The region consists of a complex system of mountain ranges that divide into three roughly parallel chains: the Siwalik Hills, Lesser Himalayas, and Great Himalayas. Several other ranges strike off from the Himalayas along the Burmese border.

In the towering, snow-covered ranges of the Great Himalayas are many of the world's highest peaks. Though the highest summits are in Nepal and China, numerous peaks in India exceed 20,000 feet (6,100 m) above sea level. Kanchenjunga, the nation's highest, reaches 28,209 feet (8,598 m). Heights of 5,000 to 15,000 feet (1,500 to 4,500 m) mark the Middle Himalayas, which, in turn, give way to low foothills of less than 4,000 feet (1,200 m) in the Outer Himalayas.

Huge glaciers and snowfields on the flanks of the higher ranges feed rivers that flow southward through deep gorges and narrow, steep-sided valleys to the Ganges Plain.

The Ganges Plain is a broad, alluvial lowland, 100 to 300 miles (160 to 480 km) wide, spanning the country south of the Himalayas. It consists mainly of the fertile basin of the Ganges River. The land is generally flat, with a slight downward slope toward the east. Much of India's farmland and many of its largest cities are on the plain, one of the most densely settled areas on earth. Only the Thar (Great Indian) Desert, an almost barren area in the west, is sparsely populated and little used.

The Deccan Plateau, often called simply the Deccan, occupies the peninsula south of the Ganges Plain. It is roughly triangular in shape and consists of a vast tableland broken by river valleys, with areas of rolling hills. The land slopes gently downward toward the east.

Fringing the Deccan on the north are the Chota Nagpur Plateau and a maze of low mountains, including the Vindhya and Satpura ranges, in the west. Elsewhere the plateau is bordered by escarpments, known as ghats. The Western Ghats, 3,000 to 5,000 feet (900 to 1,500 m) high, form a sheer wall that drops abruptly to a narrow coastal plain along the Arabian Sea.

The Eastern Ghats, in contrast, consist of low, disconnected ranges that slope gently toward a broader coastal plain along the Bay of Bengal. At the southern end of the peninsula, connecting the Eastern and Western Ghats, are the Nilgiri, Anaimalai, and Cardamom hills.

Water

India's principal river is the Ganges. It flows from the western Himalayas to the Bay of Bengal—a distance of more than 1,500 miles (2,400 km). Together with such tributaries as the Yamuna, Son, Ghaghara, and Gandak, the Ganges drains most of the mountains and the plains and part of the Deccan. In the northeast the Brahmaputra River follows a tortuous course through the Assam Valley from China and merges with the Ganges to form a vast delta, lying partly in India and partly in Bangladesh. The Ravi and Sutlej rivers, part of the Indus system, flow across northern India to Pakistan.

The Deccan is drained primarily by the eastward-flowing Mahanadi, Godavari, Krishna, and Cauvery rivers. They form large, fertile deltas at their mouths on the Bay of Bengal. The Narmada and Tapti are the only sizable rivers flowing to the Arabian Sea.

India's rivers are used extensively for irrigation; they are also used for hydroelectric power and navigation. Of particular importance is the Indus River system, which provides water for the dry Punjab region of India and Pakistan. The Indus Waters Treaty of 1960 allocates the waters of the Ravi and Sutlej to India and the waters of the Indus and its other major tributaries to Pakistan. Bhakra Dam, on the Sutlej River, is the largest of many dams in India and one of the highest dams in the world.

Climate

Sheltered by the Himalayas from the climatic extremes of Asia's interior, most of India has a tropical or subtropical climate, strongly influenced by the monsoonal wind system of southern Asia. In India the monsoons are characterized by an outward flow of relatively cool, dry air from central Asia during winter and a reverse flow of warm, moist air from the sea in summer. Other factors—such as latitude, elevation, nearness to the oceans, and location on the windward or leeward side of mountains—help determine the climate of any given area. In the Himalayas, for example, climate varies from humid subtropical in the eastern foothills to perpetually cold in the highest ranges.

Three seasons are generally recognized in India—the cool, the hot, and the rainy.

The cool season lasts from October or November until early March. Average temperatures in December and January, the coolest months, vary from about 55° F. (13° C.) on the northern edge of the Ganges Plain to between 70° and 80° F. (21° and 27° C.) in the coastal cities of the south. For most of the country this is the dry season. A notable exception is the southeastern coast, which receives much of its annual rainfall with the winter, or northeast, monsoon in October, November, and December. There is heavy snowfall in the Himalayas.

The hot season prevails from about mid-March until June; it is extremely dry. Temperatures rise rapidly over most of the country, reaching averages of 85° to 95° F. (29° to 35° C.) in May, the hottest month. Daytime highs often exceed 100° F. (38° C.), especially on the Ganges Plain. Nights bring only slight relief from the intense heat.

The rainy season begins in June with the onset of the summer, or southwest, monmys and mongooses are among the most numerous of India's many small mammals. In some regions, the monkeys, seldom killed because they have religious significance to Hindus, have become serious pests, destroying valuable food crops. Snakes, crocodiles, and other reptiles abound in India. The poisonous cobra and krait, two of the world's deadliest snakes, cause thousands of deaths each year.

Physical Features

Himalayan Mountain

The Himalayas and the associated mountains arcs gridling the sub continent on the stretch in a consistent north west- south east direction for about 2400 km between the gorges of the indus and the Tsango-Bhramaputra.The section between the Indus and the Sutlej and the Kali is termed as Kumaon Himalayas. The other two sections between the Kali and the Tista and between the latter river and the Dihangare described as the Nepal and the Assam Himalayas. Kanchanjunga (8598 Mtrs) is the highest mountain peak in India.The Greater Himalayas which have an average altitude of 6000 m have within them almost all the prominent peaks such the Everest (8848m) , Kanchenjunga (8598m) Nanga Parbat (8126m) ,Nanda devi (7817m) and Namcha parbat (7756m).

The Indus-Ganga-Brahmaputra Plain

The great plain of india is formed by the Indus, ganga and the Brahmaputra rivers. the plain extends for 3200 km between the mouths of the Ganga and the indus, all along the foot of the mountain rim, with a width varying from 150 to 300 km. The longitudinal extent from the banks of the Ravi and the Sutlej to the ganga delta alone is of 2400km. The plain is narrowest in Assam and broadens towards the west . It is 160 km wide near the Rajmahal Hills and 280 km near Allahabad. The plains are alluvial in nature.

Peninsular Plateau

Rising from the alluvial plains of uttar pradesh and Bihar, south of the Yamuna Ganga line, the great indian plateau extends towards the south to encompass the whole of Peninsula. With a general elevation of 600-900m,the plateau makes an irregular tringale with its concave base lying between Delhi ridge and Rajmahal hills and the Apex formed by Kanya Kumari . The outlying projections of the peninsular plateau presented by the Aravallis, Rajmahal and Shillong hills convey some idea of its original northerly limits.

The location of another fragment of the peninsular block in the Shillong plateau gives the indication of the possible connection. The Shillong Plateau a highly dissected and jungly tract, descends in a deep slope towards the Surma valley. The northern outliers are represented by the Mikir and the Rengma hills.

Western Ghats

The topography of the Deccan and the Karnataka Plateau is dominated by the Western Ghats, which stretch uninterruptedly to the southern tip of Peninsula.They have a general altitude of 900-1100 m but occassionally rise upto 1600 m or even more. Near Goa the highly dissected relief of the lava rocks is replaced by smoothly rounded hills of Granite and Gnesis.In this stretch the ghats dip but rise once again in the Nilgiris.Further south the continuity of the ghats is distributed by the palghat gap and the Shencottah gap.The Cardamom Hills may be regarded as the continuation of the Western ghats.

Eastern Ghats

The eastern Ghats are generally less impressive than theWestern Ghats and form a discontinous crest on the eastern peripheryof the plateau.They are repesented by an irregular line of hills, such as the Nallamalais, Velikondas,Palkondas and the Pachaimalais. This hills are often referred to as the northern hills in the northern sector,Cuddapah ranges in the middle and the Tamil nadu hills in the south.

Coastal Plains and the Islands

The plateau is flanked by coastal plains of varid width extending from Kutch to Orissa. There are striking difference between the eastern and the western coastal plains; with notable exception of Gujarat the west coast has narrow alluvial margin interspersed by hillty terrain .It has indentation except in the south where the beautiful Lagoons introduce an element of diversity.

The eastern coast on the other hand has a wide plain with well developed deltas of the major rivers. The climatic transition between the south west monsoon regime of the north and the north -east monsoon regime of the south has given rise to interesting differences in the alluvial features in the two different stretches of the east coastal plain.

River System of India

The rivers may be classified as follows :

- The Himalayanb.
- The Deccanc.
- The Coastald.
- The rivers of the inland drainage basin

The Himalayan Rivers

The Himalayan rivers are generally snow-fed and flow throughout the year. During the monsoon months (June to September), the Himalayas receive very heavy rainfall and the rivers carry the maximum amount of water, causing frequent floods.

The Deccan Rivers

The Deccan rivers are generally rain-fed and, therefore, fluctuate greatly in volume. A very large number of them are non-perennial.

The Coastal Rivers

The coastal rivers, especially on the west coast, are short and have limited catchment areas. Most of these are non-perennial as well. The rivers on the inland drainage basin are few and ephemeral.

The Rivers of the Inland Drainage Basin

They drain towards individual basins or salt lakes like the Sambhar or are lost in the sands, having no outlet to the sea.

Others Geographical Facts

Climate

The Himalayan range in the north acts as the perfect meteorological barrier for the whole country.Even though it falls under Monsoon climatical category, the climate varies from one place to another.Also despite the country's size and its varied relief, the seasonal rhythm of the monsoon is apparent throughout.Some mountains are adjoined to the north by China Nepal and Bhutan.

A series of mountain ranges separate India from Burma.Although much of northern India lies beyond the tropical zone, the entire country has a tropical climate marked by relatively high temperatures and dry winters.

Rainfall is very heavy in the north-eastern region, the western slopes of the Western Ghats and parts of the Himalayas, all of which receive over 2,000 mm annually.

The eastern part of the peninsula, extending up to the northern plains, receives rainfall varying from 1,000 to 2,000 mm a year, while the area from Western Deccan up to the Punjab plains gets between 100 mm and 500 mm a year. Rajasthan Kachchh and Ladakh have hardly any rainfall.The population of India crossed the billion mark at the turn of the millenium.

Natural Vegetation

The Himalayan region, which is rich in vegetative life, possesses varieties that can be found practically from the tropical to tundra regions. Only the altitude influences the distribution of vegetation. In the rest of the country, the type of vegetation is largely determined by the amount of rainfall. Outside the Himalayan region, the country can be divided into three major vegetation regions: the tropical wet evergreen and semi-evergreen forests, the tropical deciduous forests, and the thorn forests and shrubs.

Vegetation of the Assam region in the east is luxuriant with evergreen forests, occasional thick clumps of bamboo and tall grasses. The Gangetic plain is largely under cultivation. The Deccan tableland supports vegetation from scrub to mixed deciduous forests. The Malabar region is rich in forest vegetation. The Andaman and Nicobar Islands have evergreen, mangrove, beach and diluvial forests. Much of the country's flora originated three million years ago and are unique to the sub-continent.

Population

The population of India crossed the billion mark at the turn of the millenium. The mammoth census of 2001 is in the process of being compiled. In 1996, the population was 945 million with 73% in rural areas. In terms of population, India is the world's second-largest country, after China. 16% of the world's population lives in India. The average population density is 320 per sq km (in 1996), though it reaches 6,888 per sq km in the larger cities. In July 2003 it was 1,049,700,118 (est.)

Languages

India, according to a recent census has 1,652 dialects. Needless to say that most of them are only spoken dialects. The principal languages with rich literary heritage are :- Assamese, Bengali, English, Gujarati, Hindi, Kannada, Kashmiri, Malayalam, Marathi, Oriya, Punjabi, Sanskrit, Sindhi, Tamil, Telugu and Urdu.

Natural Resources

Coal (Fourth-largest reserves in the world), Iron ore, Manganese, Mica, Bauxite, Titanium ore, Chromite, Natural gas, Diamonds, Petroleum, Limestone, Arable land.

Environment-current Issues

Deforestation, Soil erosion, Overgrazing, Desertification, Air pollution from industrial effluents and vehicle emissions, Water pollution from raw sewage and runoff of agricultural pesticides, Tap water is not potable throughout the country, Huge and growing population is overstraining natural resources.

India is a land of incredible diversity. The diversity of the Indian population is matched by the incredible physical diversity. The sixteen official languages of India, the five major religions, and the caste system create somewhat chaotic conditions for the nation. It dominates South Asian subcontinent; near important Indian Ocean trade routes.

Sub-fields of Physical Geography

Physical Geography can be divided into several sub-fields, as follows:

Geomorphology is the field concerned with understanding the surface of the Earth and the processes by which it is shaped, both at the present as well as in the past. Geomorphology as a field has several sub-fields that deal with the specific landforms of various environments e.g. desert geomorphology and fluvial geomorphology, however, these sub-fields are united by the core processes which cause them; mainly tectonic or climatic processes.

Geomorphology seeks to understand landform history and dynamics, and predict future changes through a combination of field observation, physical experiment, and numerical modeling (Geomorphometry). Early studies in geomorphology are the foundation for pedology, one of two main branches of soil science.

Hydrology is predominantly concerned with the amounts and quality of water moving and accumulating on the land surface and in the soils and rocks near the surface and is typified by the hydrological cycle. Thus the field encompasses water in rivers, lakes, aquifers and to an extent glaciers, in which the field examines the process and dynamics involved in these bodies of water. Hydrology has historically had an important connection with engineering and has thus developed a largely quantitative method in its research; however, it does have an earth science side that embraces the systems approach.

Similar to most fields of physical geography it has sub-fields that examine the specific bodies of water or their interaction with other spheres e.g. limnology and ecohydrology.

Glaciology is the study of glaciers and ice sheets, or more commonly the cryosphere or ice and phenomena that involve ice. Glaciology groups the latter (ice sheets) as continental glaciers and the former (glaciers) as alpine glaciers. Although, research in the areas are similar with research undertaken into both the dynamics of ice sheets and glaciers the former tends to be concerned with the interaction of ice sheets with the present climate and the latter with the impact of glaciers on the landscape. Glaciology also has a vast array of sub-fields examining the factors and processes involved in ice sheets and glaciers e.g. snowhydrology and glacial geology.

Biogeography is the science which deals with geographic patterns of species distribution and the processes that result in these patterns. Biogeography emerged as a field of study as a result of the work of Alfred Russel Wallace, although the field prior to the late twentieth century had largely been viewed as historic in its outlook and descriptive in its approach. The main stimulus for the field since its founding has been that of evolution, plate tectonics and the theory of island biogeography. The field can largely be divided into five sub-fields: island biogeography, paleobiogeography, phylogeography, zoogeography and phytogeography.

Climatology is the study of the climate, scientifically defined as weather conditions averaged over a long period of time. Climatology examines both the nature of micro (local) and macro (global) climates and the natural and anthropogenic influences on them. The field is also sub-divided largely into the climates of various regions and the study of specific phenomena or time periods e.g. tropical cyclone rainfall climatology and paleoclimatology.

Meteorology is the interdisciplinary scientific study of the atmosphere that focuses on weather processes and short term forecasting (in contrast with climatology). Studies in the field stretch back millennia, though significant progress in meteorology did not occur until the eighteenth century. Meteorological phenomena are observable weather events which illuminate and are explained by the science of meteorology.

Pedology is the study of soils in their natural environment. It is one of two main branches of soil science, the other being edaphology. Pedology mainly deals with pedogenesis, soil morphology, soil classification. In physical geography pedology is largely studied due to the numerous interactions between climate (water, air, temperature), soil life (micro-organisms, plants, animals), the mineral materials within soils (biogeochemical cycles) and its position and effects on the landscape such as laterization.

Palaeogeography is a cross-disciplinary study that examines the preserved material in the stratigraphic record in order to determine the distribution of the continents through geologic time. Almost all the evidence for the positions of the continents comes from geology in the form of fossils or paleomagnetism. The use of this data has resulted in evidence for continental drift, plate tectonics and supercontinents. This in turn has supported palaeogeographic theories such as the Wilson cycle.

Coastal geography is the study of the dynamic interface between the ocean and the land, incorporating both the physical geography (i.e. coastal geomorphology, geology and oceanography) and the human geography of the coast. It involves an understanding of coastal weathering processes, particularly wave action, sediment movement and weathering, and also the ways in which humans interact with the coast. Coastal geography although predominantly geomorphological in its research is not just concerned with coastal landforms, but also the causes and influences of sea level change.

Oceanography is the branch of physical geography that studies the Earth's oceans and seas. It covers a wide range of topics, including marine organisms and ecosystem dynamics (biological oceanography); ocean currents, waves, and geophysical fluid dynamics (physical oceanography); plate tectonics and the geology of the sea floor (geological oceanography); and fluxes of various chemical substances and physical properties within the ocean and across its boundaries (chemical oceanography).

These diverse topics reflect multiple disciplines that oceanographers blend to further knowledge of the world ocean and understanding of processes within it.

Quaternary science is an inter-disciplinary field of study focusing on the Quaternary period, which encompasses the last 2.6 million years. The field studies the last ice age and the recent interstadial the Holocene and uses proxy evidence to reconstruct the past environments during this period to infer the climatic and environmental changes that have occurred.

Landscape ecology is a sub-discipline of ecology and geography that address how spatial variation in the landscape affects ecological processes such as the distribution and flow of energy, materials and individuals in the environment (which, in turn, may influence the distribution of landscape "elements" themselves such as hedgerows). The field was largely founded by the German geographer Carl Troll Landscape ecology typically deals with problems in an applied and holistic context. The main difference between biogeography and landscape ecology is that the latter is concerned with how flows or energy and material are changed and their impacts on the landscape whereas the former is concerned with the spatial patterns of species and chemical cycles.

Geomatics is the field of gathering, storing, processing, and delivering of geographic information, or spatially referenced information. Geomatics includes geodesy (scientific discipline that deals with the measurement and representation of the earth, its gravitational field, and other geodynamic phenomena, such as crustal motion, oceanic tides, and polar motion) and GIS (a computer based system for capturing, storing, analyzing and managing data and associated attributes which are spatially referenced to the earth) and remote sensing (the short or large-scale acquisition of information of an object or phenomenon, by the use of either recording or real-time sensing devices that are not in physical or intimate contact with the object).

Environmental geography is a branch of geography that analyzes the spatial aspects of interactions between humans and the natural world. The branch bridges the divide between human and physical geography and thus requires an understanding of the dynamics of geology, meteorology, hydrology, biogeography, and geomorphology, as well as the ways in which human societies conceptualize the environment. Although the branch was previously more visible in research than at present with theories such as environmental determinism linking society with the environment. It has largely become the domain of the study of environmental management or anthropogenic influences.

Origin and Development

From the birth of geography as a science during the Greek classical period and until the late nineteenth century with the birth of anthropogeography or Human Geography, Geography was almost exclusively a natural science: the study of location and descriptive gazetteer of all places of the known world. Several works among the best known during this long period could be cited as an example, from Strabo (Geography), Eratosthenes (Geography) or Dionisio Periegetes (Periegesis Oiceumene) in the Ancient Age to the Alexander von Humboldt (Cosmos) in the century XIX, in which geography is regarded as a physical and natural science, of course, through the work Summa de Geografía of Martín Fernández de Enciso from the early sixteenth century, which is indicated for the first time the New World.

During the eighteenth and nineteenth centuries, a controversy exported from Geology, between supporters of James Hutton (uniformitarianism Thesis) and Georges Cuvier (catastrophism) strongly influenced the field of geography, because geography at this time was a natural science since Human Geography or Antropogeography had just developed as a discipline in the late nineteenth century.

Two historical events during the nineteenth century had a great effect in the further development of physical geography. The first was the European colonial expansion in Asia, Africa, Australia and even America in search of raw materials required by industries during the Industrial Revolution. This fostered the creation of geography departments in the universities of the colonial powers and the birth and development of national geographical societies, thus giving rise to the process identified by Horacio Capel as the institutionalization of geography.

One of the most prolific empires in this regard was the Russian. A mid-eighteenth century many geographers are sent by the Russian altamirazgo different opportunities to perform geographical surveys in the area of Arctic Siberia. Among these is who is considered the patriarch of Russian geography: Mikhail Lomonosov who in the mid-1750s began working in the Department of Geography, Academy of Sciences to conduct research in Siberia, their contributions are notable in this regard, shows the soil organic origin, develops a comprehensive law on the movement of the ice that still governs the basics, thereby founding a new branch of Geography:

Glaciology. In 1755 his initiative was founded Moscow University where he promotes the study of geography and the training of geographers. In 1758 he was appointed director of the Department of Geography, Academy of Sciences, a post from which would develop a working methodology for geographical survey guided by the most important long expeditions and geographical studies in Russia.

Thus followed the line of Lomonosov and the contributions of the Russian school became more frequent through his disciples, and in the nineteenth century we have great geographers as Vasily Dokuchaev who performed works of great importance as a "principle of comprehensive analysis of the territory" and "Russian Chernozem" latter being the most important where introduces the geographical concept of soil, as distinct from a simple geological strata, and thus founding a new geographic area of study: the Pedology.

Climatology also receive a strong boost from the Russian school by Wladimir Köppen whose main contribution, climate classification, is still valid today. However, this great geographer also contributed to the Paleogeography through his work "The climates of the geological past" which is considered the father of Paleoclimatology. Russian geographers who made great contributions to the discipline in this period were: NM Sibirtsev, Pyotr Semyonov, K. D. Glinka, Neustrayev, among others.

The second important process is the theory of evolution by Darwin in mid-century (which decisively influenced the work of Ratzel, who had academic training as a zoologist and was a follower of Darwin's ideas) which meant an important impetus in the development of Biogeography.

Another major event in the late nineteenth and early twentieth century will give a major boost to development of geography and will take place in United States. It is the work of the famous geographer William Morris Davis who not only made important contributions to the establishment of discipline in his country, but revolutionized the field to develop geographical cycle theory which he proposed as a paradigm for Geography in general, although in actually served as a paradigm for Physical Geography.

His theory explained that mountains and other landforms are shaped by the influence of a number of factors that are manifested in the geographical cycle. He explained that the cycle begins with the lifting of the relief by geological processes (faults, volcanism, tectonic upheaval, etc.).. Geographical factors such as rivers and runoff begins to create the V-shaped valleys between the mountains (the stage called "youth"). During this first stage, the terrain is steeper and more irregular.

Over time, the currents can carve wider valleys ("maturity") and then start to wind, towering hills only ("senescence"). Finally, everything comes to what is a plain flat plain at the lowest elevation possible (called "baseline") This plain was called by Davis' "peneplain" meaning "almost plain" Then the rejuvenation occurs and there is another mountain lift and the cycle continues. Although Davis's theory is not entirely accurate, it was absolutely revolutionary and unique in its time and helped to modernize and create Geography subfield of Geomorphology. Its implications prompted a myriad of research in various branches of Physical Geography. In the case of the Paleogeography this theory provided a model for understanding the evolution of the landscape.

For Hydrology, Glaciology and Climatology as a boost investigated as studying geographic factors shape the landscape and affect the cycle. The bulk of the work of William Morris Davis led to the development of a new branch of Physical Geography: Geomorphology whose contents until then did not differ from the rest of Geography. Shortly after this branch would present a major development. Some of his disciples made significant contributions to various branches of physical geography such as Curtis Marbut and his invaluable legacy for Pedology, Mark Jefferson, Isaiah Bowman, among others.

Atmosphere

It is a layer of gases that may surround a material body of sufficient mass, and that is held in place by the gravity of the body. An atmosphere may be retained for a longer duration, if the gravity is high and the atmosphere's temperature is low. Some planets consist mainly of various gases, but only their outer layer is their atmosphere.

The term stellar atmosphere describes the outer region of a star, and typically includes the portion starting from the opaque photosphere outwards. Relatively low-temperature stars may form compound molecules in their outer atmosphere. Earth's atmosphere, which contains oxygen used by most organisms for respiration and carbon dioxide used by plants, algae and cyano-bacteria for photosynthesis, also protects living organisms from genetic damage by solar ultraviolet radiation. Its current composition is the product of billions of years of biochemical modification of the paleoatmosphere by living organisms.

Atmospheric Pressure

Atmospheric pressure is just another term for air pressure. One way to explain air pressure is to think of it as the force the atmosphere places on a specified area. The pressure can be measured by a barometer.

How to Calculate Air Pressure

The amount of pressure will depend on several factors, but the primary one is altitude. You can calculate the pressure with a scientific calculator.

To do this, you need to define the physical constraints of the planet. The universal gas constant R has a value of 8.31432 Newton meters/moles Kelvin. The typical gravity exerted by the planet at sea level is 9.80665 m per second squared (m/s^2). The molar mass is M of the air 0.0289644 kilograms per mole (kg/mol).

Define the Height

Next, you have to determine the height. H is the measurement over sea level. HH is the minimum zone altitude the measurement point is at. The minimum altitudes are in the thousandths of meters: 0, 11, 20, 32, 47, 51 and 71.

Define the Temperature

The standard temperature values (Tb) will be based on the altitude zones provided (in Kelvin): 288.15, 216.65, 216.65, 228.65, 270.65, 270.65 and 214.65. The typical temperature lapse rate (Lb) decreases at the specified latitude.

In this example of how to measure atmospheric pressure, the lapse rates (Kelvin per meter) are as follows: -.0065, 0, .001, .0028, 0 -.0028 and -.002.

Define the Pressure and Calculate

Here, the altitude zone pressure is in pascals: 101,325, 22,632, 5474, 868, 110, 66 and 4. To calculate the pressure (P), use the formula P = Pb [Tb/(Tb + Lb (h - hb))]^(goM/RLb).

Air Pressure and its Effects on People

Air pressure does not just affect the environment; it also has a direct effect on people, particularly blood pressure. The air pressure changes underwater, when deep sea diving or departing from the atmosphere.

Depending on the altitude, the air pressure changes and this affects the blood pressure. When deep sea diving, the pressure increases greatly. The water generates a lot of pressure on the swimmer. This can boost blood pressure. This can be dangerous for people who have high blood pressure.

A different effect is experienced by astronauts. The effect is very different because there is no gravity. When they get back to Earth, they may become susceptible to fainting as the blood pressure goes down.

Gravity

Gravity affects the atmosphere because the molecules in the air are pulled by gravitational forces. The pressure on the molecules will hinge on the amount of gravity present. For instance, there is less barometric pressure on the Moon than on the Earth because the gravity there is lesser.

Density

Mass air density influences pressure. If the air is dense, the higher the barometric will be, and vice versa.

The atmospheric pressure is essential for life on Earth to flourish. As the facts show, it has an effect on people and other living organisms. An understanding of how it works can help you get insights on why it is so vital.

Atmospheric Escape

Surface gravity, the force that holds down an atmosphere, differs significantly among the planets. For example, the large gravitational force of the giant planet Jupiter is able to retain light gases such as hydrogen and helium that escape from lower gravity objects. Second, the distance from the sun determines the energy available to heat atmospheric gas to the point where its molecules' thermal motion exceed the planet's escape velocity the speed at which gas molecules overcome a planet's gravitational grasp.

Thus the distant and cold Titan,Triton, and Pluto are able to retain their atmospheres despite relatively low gravities. Interstellar planets, theoretically, may also retain thick atmospheres.

Since a gas at any particular temperature will have molecules moving at a wide range of velocities, there will almost always be some slow leakage of gas into space. Lighter molecules move faster than heavier ones with the same thermal kinetic energy, and so gases of low molecular weight are lost more rapidly than those of high molecular weight. It is thought that Venus and Mars may have both lost much of their water when, after being photo dissociated into hydrogen and oxygen by solar ultraviolet, the hydrogen escaped.

Earth's magnetic field helps to prevent this, as, normally, the solar wind would greatly enhance the escape of hydrogen. However, over the past 3 billion years the Earth may have lost gases through the magnetic polar regions due to auroral activity, including a net 2% of its atmospheric oxygen.

Other mechanisms that can cause atmosphere depletion are solar wind-induced sputtering, impact erosion, weathering, and sequestration — sometimes referred to as "freezing out" — into the regolith and polar caps.

Composition

Earth, our home, is the third planet from the sun. It is the only planet known to have an atmosphere containing free oxygen, oceans of liquid water on its surface, and, of course, life.

Earth is the fifth largest of the planets in the solar system — smaller than the four gas giants, Jupiter, Saturn, Uranus and Neptune, but larger than the three other rocky planets, Mercury, Marsand Venus.

It has a diameter of roughly 8,000 miles (13,000 kilometers), and is round because gravity pulls matter into a ball, although it is not perfectly round, instead being more of an "oblate spheroid" whose spin causes it to be squashed at its poles and swollen at the equator. Roughly 71 percent of Earth's surface is covered by water, most of it in the oceans. About a fifth of its atmosphere is made up of oxygen, produced by plants.

Initial atmospheric makeup is generally related to the chemistry and temperature of the local solar nebula during planetary formation and the subsequent escape of interior gases. These original atmospheres underwent much evolution over time, with the varying properties of each planet resulting in very different outcomes.

The atmospheres of the planets Venus and Mars are primarily composed of carbon dioxide, with small quantities of nitrogen, argon, oxygen and traces of other gases.

The atmospheric composition on Earth is largely governed by the by-products of the very life that it sustains. Earth's atmospherecontains roughly (by molar content/volume) 78.08% nitrogen, 20.95% oxygen, a variable amount (average around 1.247%, National Center for Atmospheric Research) water vapor, 0.93% argon, 0.038% carbon dioxide, and traces of hydrogen, helium, and other "noble" gases.

The low temperatures and higher gravity of the gas giants — Jupiter, Saturn, Uranus and Neptune — allows them to more readily retain gases with low molecular masses. These planets have hydrogen-helium atmospheres, with trace amounts of more complex compounds.

Two satellites of the outer planets possess non-negligible atmospheres: Titan, a moon of Saturn, and Triton, a moon of Neptune, which are mainly nitrogen. Pluto, in the nearer part of its orbit, has an atmosphere of nitrogen and methane similar to Triton's, but these gases are frozen when farther from the Sun.

Other bodies within the Solar System have extremely thin atmospheres not in equilibrium. These include the Moon (sodium gas), Mercury (sodium gas), Europa (oxygen), Io (sulfur), andEnceladus (water vapor).

The atmospheric composition of an extra-solar planet was first determined using the Hubble Space Telescope. Planet HD 209458b is a gas giant with a close orbit around a star in theconstellation Pegasus. The atmosphere is heated to temperatures over 1,000 K, and is steadily escaping into space. Hydrogen, oxygen, carbon and sulfur have been detected in the planet's inflated atmosphere.

Orbital Characteristics

The Earth spins on an imaginary line called an axis that runs from the north pole to the south pole, while also orbiting the sun. It takes Earth 24 hours to complete a rotation on its axis, and roughly 365 days to complete an orbit around the sun.

The Earth's axis of rotation is tilted in relation to the ecliptic plane, an imaginary surface through Earth's orbit around the sun. This means the northern and southern hemispheres will sometimes point toward or away from the sun depending on the time of year, varying the amount of light they receive and causing the seasons.

Earth's orbit is not a perfect circle, but is rather an oval-shaped ellipse, like that of the orbits of all the other planets. Earth is a bit closer to the sun in early January and farther away in July, although this variation has amuch smaller effect than the heating and cooling caused by the tilt of Earth's axis. Earth happens to lie within the so-called "Goldilocks zone" around its star, where temperatures are just right to maintain liquid water on its surface.

History

Earth probably formed at roughly the same time as the sun and other planets some 4.6 billion years ago, when the solar system coalesced from a giant, rotating cloud of gas and dust known as the solar nebula. As the nebula collapsed because of its gravity, it spun faster and flattened into a disk. Most of the material was pulled toward the center to form the sun. Other particles within the disk collided and stuck together to form ever-larger bodies, including the Earth. The solar wind from the sun was so powerful that it swept away most of the lighter elements, such as hydrogen and helium, from the innermost worlds, rendering Earth and its siblings into small, rocky planets.

Scientists think Earth started off as a waterless mass of rock. Radioactive materials in the rock and increasing pressure deep within the Earth generated enough heat to melt Earth's interior, causing some chemicals to rise to the surface and form water, while others became the gases of the atmosphere. Recent evidence suggests that Earth's crust and oceans may have formed within about 200 million years after the planet had taken shape.

The history of Earth is divided into four eons — starting with the earliest, these are the Hadean, Archean, Proterozoic, and Phanerozoic. The first three eons, which together lasted nearly 4 billion years, are together known as the Precambrian. Evidence for life has bee found in the Archaean about 3.8 billion years ago, but life did not become abundant until the Phanerozoic.

The Phanerozoic is divided into three eras — starting with the earliest, these are the Paleozoic, Mesozoic, and Cenozoic. The Paleozoic Era saw the development of many kinds of animals and plants in the seas and on land, the Mesozoic Era was the age of dinosaurs, and the Cenozoic Era we are in currently is the age of mammals.

Most of the fossils seen in Paleozoic rocks are invertebrate animals lacking backbones, such as corals, mollusks and trilobites. Fish are first found about 450 million years ago, while amphibians appear roughly 380 million years ago. By 300 million years ago, large forests and swamps covered the land, and the earliest fossils of reptiles appear during this period as well.

The Mesozoic saw the ascendence of dinosaurs, although mammals also appear in the fossil record about 200 million years ago. During this time, flowering plants became the dominant plant group and continue to be so today.

The Cenozoic began about 65 million years ago with the end of the age of dinosaurs, which many scientists think was caused by a cosmic impact. Mammals survived to become the dominant land animals of today.

Composition & Structure

Atmosphere

The atmosphere is roughly 78 percent nitrogen, 21 percent oxygen, with trace amounts of water, argon, carbon dioxide and other gases. Nowhere else in the solar system can one find an atmosphere loaded with free oxygen, which ultimately proved vital to one of the other unique features of Earth — us.

Air surrounds Earth and becomes thinner farther from the surface. Roughly 100 miles (160 kilometers) above Earth, the air is so thin that satellites can zip through with little resistance. Still, traces of atmosphere can be found as high as 370 miles (600 kilometers) above the surface.

The lowest layer of the atmosphere is known as the troposphere, which is constantly in motion, causing the weather. Sunlight heats the Earth's surface, causing warm air to rise. This air ultimately expands and cools as air pressure decreases, and because this cool air is denser than its surroundings, it then sinks, only to get warmed by the Earth once again.

Above the troposphere, some 30 miles (48 kilometers) above the Earth's surface, is the stratosphere. The still air of the stratosphere contains the ozone layer, which was created when ultraviolet light caused trios of oxygen atoms to bind together into ozone molecules. Ozone prevents most of the sun's harmful ultraviolet radiation from reaching Earth's surface.

Water vapor, carbon dioxide and other gases in the atmosphere trap heat from the sun, warming Earth. Without this so-called "greenhouse effect," Earth would probably be too cold for life to exist, although a runaway greenhouse effect led to the hellish conditions now seen on Venus.

Earth-orbiting satellites have shown that the upper atmosphere actually expands during the day and contracts at night due to heating and cooling.

Magnetic field

The Earth's magnetic field is generated by currents flowing in Earth's outer core. The magnetic poles arealways on the move, with the magnetic north pole recently accelerating its northward motion to 24 miles (40 km) annually, likely exiting North America and reaching Siberia in a few decades.

Earth's magnetic field is changing in other ways, too — globally, the magnetic field has weakened 10 percent since the 19th century. These changes are mild compared to what Earth's magnetic field has done in the past — sometimes the field completely flips, with the north and the south poles swapping places.

When charged particles from the sun get trapped in Earth's magnetic field, they smash into air molecules above the magnetic poles, causing them to glow, a phenomenon known as the aurorae, the northern and southern lights.

Chemical Composition

Oxygen is the most abundant element in rocks in Earth's crust, composing roughly 47 percent of the weight of all rock. The second most abundant element is silicon at 27 percent, followed by aluminum at 8 percent, iron at 5 percent, calcium at 4 percent, and sodium, potassium, and magnesium at about 2 percent each.

The Earth's core consists mostly of iron and nickel and potentially smaller amounts of lighter elements such as sulfur and oxygen. The mantle is made of iron and magnesium-rich silicate rocks. (The combination of silicon and oxygen is known as silica, and minerals that contain silica are known as silicate minerals.)

Internal Structure

The Earth's core is about 4,400 miles (7,100 kilometers) wide, slightly larger than half the Earth's diameterand roughly the size of Mars. The outermost 1,400 miles (2,250 kilometers) of the core are liquid, while the inner core — about four-fifths as big as Earth's moon at some 1,600 miles (2,600 kilometers) in diameter — is solid.

Above the core is Earth's mantle, which is about 1,800 miles (2,900 kilometers) thick. The mantle is not completely stiff, but can flow slowly. Earth's crust floats on the mantle much as a wood floats on water, and the slow motion of rock in the mantle shuffles continents around and causes earthquakes, volcanoes, and the formation of mountain ranges.

Above the mantle, Earth has two kinds of crust. The dry land of the continents consists mostly of granite and other light silicate minerals, while the ocean floors are made up mostly of a dark, dense volcanic rock called basalt. Continental crust averages some 25 miles (40 kilometers) thick, although it can be thinner or thicker in some areas. Oceanic crust is usually only about 5 miles (8 kilometers) thick. Water fills in low areas of the basalt crust to form the world's oceans. Earth has more than enough water to completely fill the ocean basins, and the rest of it spreads onto edges of the continents, areas known as the continental shelf.

Earth gets warmer toward its core. At the bottom of the continental crust, temperatures reach about 1,800 degrees F (1,000 degrees C), increasing about 3 degrees F per mile (1 degrees C per kilometer) below the crust. Geologists think the temperature of Earth's outer core is about 6,700 to 7,800 degrees F (3,700 to 4,300 degrees C), and the inner core may reach 12,600 degrees F (7,000 degrees C), hotter than the surface of the sun. Only the enormous pressures found at the super-hot inner core keep it solid.

Structure

Earth's Atmosphere

The atmosphere is a mixture of nitrogen (78%), oxygen (21%), and other gases (1%) that surrounds Earth. High above the planet, the atmosphere becomes thinner until it gradually reaches space. It is divided into five layers. Most of the weather and clouds are found in the first layer.

The atmosphere is an important part of what makes Earth livable. It blocks some of the Sun's dangerous rays from reaching Earth. It traps heat, making Earth a comfortable temperature. And the oxygen within our atmosphere is essential for life.

Over the past century, greenhouse gases and other air pollutants released into the atmosphere have been causing big changes like global warming, ozone holes, and acid rain.

The Earth's atmosphere consists, from the ground up, of the troposphere (which includes the planetary boundary layer or peplosphere as lowest layer), stratosphere (which includes theozone layer), mesosphere, thermosphere (which contains the ionosphere), exosphere and also the magnetosphere. Each of the layers has a different lapse rate, defining the rate of change in temperature with height.

can transport thermal energy from the higher temperature interior up to the surface.

Importance

From the perspective of the planetary geologist, the atmosphere is an evolutionary agent essential to the morphology of a planet. The wind transports dust and other particles which erodes the relief and leaves deposits (eolian processes). Frost and precipitations, which depend on the composition, also influence the relief. Climate changes can influence a planet's geological history. Conversely, studying surface of earth leads to an understanding of the atmosphere and climate of a planet - both its present state and its past.

For a meteorologist, the composition of the atmosphere determines the climate and its variations. For a biologist, the composition is closely dependent on the appearance of the life and its evolution.

Biosphere

The biosphere is the global sum of all ecosystems. It can also be called the zone of life on Earth, a closed (apart from solar and cosmic radiation) and self-regulating system. From the broadest biophysiological point of view, the biosphere is the global ecological system integrating all living beings and their relationships, including their interaction with the elements of the lithosphere, hydrosphere and atmosphere.

The biosphere is postulated to have evolved, beginning through a process of biogenesis or biopoesis, at least some 3.5 billion years ago. In a broader sense; biospheres are any closed, self-regulating systems containing ecosystems; including artificial ones such as Biosphere 2 and BIOS-3; and, potentially, ones on other planets or moons.

The term "biosphere" was coined by geologist Eduard Suess in 1875, which he defined as:

> "The place on Earth's surface where life dwells."

While this concept has a geological origin, it is an indication of the impact of both Darwin and Maury, [Nicholas Morken on the earth sciences.

The biosphere's ecological context comes from the 1920s, preceding the 1935 introduction of the term "ecosystem" by Sir Arthur Tansley. Vernadsky defined ecology as the science of the biosphere. It is an interdisciplinary concept for integrating astronomy, geophysics, meteorology, biogeography, evolution, geology, geochemistry, hydrology and, generally speaking, all life and earth sciences.

Narrow Definition

Some life scientists and earth scientists use biosphere in a more limited sense. For example, geochemists define the biosphere as being the total sum of living organisms (the "biomass" or "biota" as referred to by biologists and ecologists). In this sense, the biosphere is but one of four separate components of the geochemical model, the other three being lithosphere, hydrosphere, and atmosphere. The narrow meaning used by geochemists is one of the consequences of specialization in modern science. Some might prefer the word ecosphere, coined in the 1960s, as all encompassing of both biological and physical components of the planet.

The Second International Conference on Closed Life Systems defined biospherics as the science and technology of analogs and models of Earth's biosphere; i.e., artificial Earth-like biospheres. Others may include the creation of artificial non-Earth biospheres—for example, human-centered biospheres or a native Martian biosphere—in the field of biospherics.

Gaia Hypothesis

In the early 1970s, Lynn Margulis, a microbiologist from the United States, added to the hypothesis, specifically noting the ties between the biosphere and other Earth systems. For example, when carbon dioxide levels increase in the atmosphere, plants grow more quickly. As their growth continues, they remove more and more carbon dioxide from the atmosphere.

Many scientists are now involved in new fields of study that examine interactions between biotic and abiotic factors in the biosphere, such as geobiology and geomicrobiology.

Ecosystems occur when communities and their physical environment work together as a system. The difference between this and a biosphere is simple, the biosphere is everything in general terms.

Extent of Earth's biosphere

Every part of the planet, from the polar ice caps to the Equator, supports life of some kind. Recent advances in microbiology have demonstrated that microbes live deep beneath the Earth's terrestrial surface, and that the total mass of microbial life in so-called "uninhabitable zones" may, in biomass, exceed all animal and plant life on the surface. The actual thickness of the biosphere on earth is difficult to measure. Birds typically fly at altitudes of 650 to 1,800 metres, and fish that live deep underwater can be found down to -8,372 metres in the Puerto Rico Trench.

There are more extreme examples for life on the planet: Rüppell's vulture has been found at altitudes of 11,300 metres; bar-headed geesemigrate at altitudes of at least 8,300 metres; yaks live at elevations between 3,200 to 5,400 metres above sea level; mountain goats live up to 3,050 metres. Herbivorous animals at these elevations depend on lichens, grasses, and herbs.

Microscopic organisms live at such extremes that, taking them into consideration puts the thickness of the biosphere much greater. Culturable microbes have been found in the Earth's upper atmosphere as high as 41 km (25 mi) (Wainwright et al.,

2003, in FEMS Microbiology Letters). It is unlikely, however, that microbes are active at such altitudes, where temperatures and air pressure are extremely low and ultraviolet radiation very high. More likely these microbes were brought into the upper atmosphere by winds or possiblyvolcanic eruptions. Barophilic marine microbes have been found at more than 10 km (6 mi) depth in the Marianas Trench (Takamia et al., 1997, in FEMS Microbiology Letters).

Microbes are not limited to the air, water or the Earth's surface. Culturable thermophilic microbes have been extracted from cores drilled more than 5 km (3 mi) into the Earth's crust in Sweden (Gold, 1992, and Szewzyk, 1994, both in PNAS), from rocks between 65-75 °C. Temperature increases with increasing depth into the Earth's crust. The speed at which the temperature increases depends on many factors, including type of crust (continental vs. oceanic), rock type, geographic location, etc. The upper known limit of temperature at which microbial life can exist is 122 °C (Methanopyrus kandleri Strain 116), and it is likely that the limit of life in the "deep biosphere" is defined by temperature rather than absolute depth.

Our biosphere is divided into a number of biomes, inhabited by broadly similar flora and fauna. On land, biomes are separated primarily by latitude. Terrestrial biomes lying within the Arctic and Antarctic Circles are relatively barren of plant and animal life while most of the more populous biomes lie near the equator. Terrestrial organisms in temperate and Arctic biomes have relatively small amounts of total biomass smaller energy budgets, and display prominent adaptations to cold including world-spanning migrations social adaptations homeothermy estivation and multiple layers of insulation.

Specific Biospheres

When the word is followed by a number, it is usually referring to a specific system or number. Thus:

- Biosphere 1, the planet Earth
- Biozsphere 2, a laboratory in Arizona which contains 3.15 acres (13,000 m^2) of closed ecosystem.
- BIOS-3, a closed ecosystem at the Institute of Biophysics in Krasnoyarsk, Siberia, in what was then the Soviet Union.
- Biosphere J (CEEF, Closed Ecology Experiment Facilities), an experiment in Japan.

Geosphere

The geosphere is considered that portion of the Earth system that includes the Earth's interior, rocks and minerals, landforms and the processes that shape the Earth's surface. The Earth itself (contrary to Christopher Columbus) is not a perfect sphere. It is what is called an oblate spheroid, with a radius of 6,357 kilometers (km) from the Earth's center to the North Pole and 6,378 km from the center to the Equator.

Prior to advanced instruments and spacecraft, 17th-century scientist Sir Isaac Newton predicted a similar shape based on the effects of the Earth's daily rotation and his studies of other planets. Geodesy (the study of the Earth's shape) is a very important science, in that it is critical for helping us understand satellite orbits, create maps and navigate on the planet using devices such as the Global Positioning System (GPS).

The term geosphere is often used to refer to the densest parts of Earth, which consist mostly of rock and regolith. The geosphere consists of the inside of the Earth or other planets or bodies.

In Aristotelian physics, the term was applied to four spherical natural places, concentrically nested around the center of the Earth, as described in the lectures Physica and Meteorologica. They were believed to explain the motions of the four terrestrial elements: Earth, Water, Air and Fire.

In modern texts and in Earth system science, geosphere refers to the solid parts of the Earth and is used along with atmosphere, hydrosphere, and biosphere to describe the systems of the Earth (the interaction of these systems with the heliosphere is sometimes listed). In that context, sometimes the term lithosphere is used instead of geosphere. However, the lithosphere only refers to the uppermost layers of the solid Earth (oceanic and continental crustal rocks and uppermost mantle).

Since space exploration began, it has been observed that the extent of the ionosphere or plasmasphere is highly variable, and often much larger than previously appreciated, at times extending to the boundaries of the Earth's magnetosphere or geomagnetosphere. This highly variable outer boundary of geogenic matter has been referred to as the "geopause," to suggest the relative scarcity of such matter beyond it, where the solar wind dominates.

The Earth's interior is arranged somewhat like a layer cake, consisting of a series of layers that change in density, mineral composition and thickness with depth. Directly below the crust is the mantle. It consists of two parts, an upper layer that is less dense and relatively brittle and a lower (much thicker) layer that is more dense and plastic (it deforms without breaking). The crust and upper mantle combined form the brittle upper layers of the Earth's interior called the lithosphere. The upper mantle is also called the asthenosphere.

The mantle makes up the largest volume of the Earth's interior. The region beneath the mantle is called the core, and consists of two parts, a liquid outer core that is around 2250 km thick and a solid inner core 1220 km thick. The core is primarily made up of iron, with a small amount of nickel. The liquid iron in the outer core is particularly important in that it is the primary source of the Earth's magnetic field. Unlike a common magnet, though, the north and south ends of our "global magnet" are not exactly situated at Earth's poles. Instead, the magnetic north pole is actually situated in northern Canada, and the magnetic south pole resides north of Antarctica and south of Australia. Another interesting feature of the magnetic poles is that their precise location moves over time.

Every few million years, even the polarity of the Earth's magnetic field reverses (called a geomagnetic reversal, where magnetic north and south "switch"). While scientists still do not fully understand why geomagnetic reversals occur, the presence of changing magnetic orientations preserved in rocks containing iron was a fundamental clue in unravelling the puzzle of Plate Tectonics. Almost all of our direct knowledge of the Earth's interior is from the upper 10 km. Our knowledge of the remaining 6,300 km is based largely on indirect evidence from seismology, laboratory studies of igneous and metamorphic rocks, computer models and meteorites.

Natural Science

The natural sciences are branches of science that seek to elucidate the rules that govern the natural world by using scientific methods. The term "natural science" is used to distinguish the subject matter from the social sciences, which apply the scientific method to study human behavior and social patterns; the humanities, which use a critical or analytical approach to study the human condition; and the formal sciencessuch as mathematics and logic, which use an a priori, as opposed to factual methodology to study formal systems.

Natural sciences are the basis for applied sciences. Together the natural and applied sciences are distinguished from the social sciences on the one hand and the humanities on the other. Though mathematics, statistics, and computer science are not considered natural sciences, for instance, they provide many tools and frameworks used within the natural sciences.

Alongside this traditional usage, the phrase natural sciences is also sometimes used more narrowly to refer to natural history. In this sense "natural sciences" may refer to the biology and perhaps also the earth sciences as distinguished from the physical sciences including astronomy physics, and chemistry.

Within the natural sciences, the term hard science is sometimes used to describe those subfields which some people view as relying on experimental, quantifiable data or the scientific method and focus on accuracy and objectivity. These usually include physics, chemistry and biology. By contrast, soft science is often used to describe the scientific fields that are more reliant on qualitative research, including the social sciences.

This discipline is the science of celestial objects and phenomena that originate outside the Earth's atmosphere. It is concerned with the evolution, physics, chemistry, meteorology, and motion of celestial objects, as well as the formation and development of the universe.

Astronomy includes the examination, study and modeling of stars, planets, comets, galaxies and the cosmos. Most of the information used by astronomers is gathered by remote observation, although some laboratory reproduction of celestial phenomenon has been performed (such as the molecular chemistry of the interstellar medium).

While the origins of the study of celestial features and phenomenon can be traced back to antiquity, the scientific methodology of this field began to develop in the middle of the 17th century. A key factor was Galileo's introduction of the telescope to examine the night sky in more detail.

The mathematical treatment of astronomy began with Newton's development of celestial mechanics and the laws of gravitation, although it was triggered by earlier work of astronomers such as Kepler. By the 19th century, astronomy had developed into a formal science, with the introduction of instruments such as the spectroscope and photography, along with much-improved telescopes and the creation of professional observatories.

Importance of Physical Geography

Physical geography is the branch of geography dealing with natural features and processes. It includes the study of lithosphere (Examples: landforms, drainage, relief and physiography), atmosphere (its composition, formation, elements and controls of weather and temperature, pressure, winds, rainfall, climatic types, etc.), hydrosphere (oceans, seas, lakes and related features with water realm) and biosphere (life forms including human being and macro-organism and their supporting mechanism, viz. food chain, ecological parameters and ecological balance). Soils are formed through the process of pedogenesis and depend upon the parent rocks, climate, biological activity, and time.

Physical Geography plays a very vital role in the very existence of human beings. Every student who studies about the planet earth should also study Physical Geography. This is because physical geography involves the study of the natural processes of the earth.

In Physical Geography, you will study all about the various elements of nature.

By study in this area, you can learn about the weather and how it's changing (and the possible results of those transforms) have an effect on people now and can help plan for the prospect.

Geography doesn't just find out whether humans can live in a definite area or not, it also determines people's lifestyles, as they adjust to the obtainable food and atmosphere patterns.

It also deals with the interaction of our planet Earth with that of the sun, moon, stars, space.

It also covers all aspects of the earth in relation to the climatic conditions, the seasons, atmospheric force and composition, the effects of the wind, storms, rain, snow, and other aspects.

It deals with the various climatic zones, flora and fauna, wild animals, the hydrological cycles, precipitation, the wind pressures, the microclimates, soil erosion, desserts, etc.

The study of physical Geography is essential for the sufficient allocation of the natural resources on the earth. It is essential for enabling human resolution as per the adjacent conditions and to be improved informed in order to protect our planet earth.

Time provides maturity to soils and facilitates in the maturity of soil shapes. Each component is significant for human beings. Landforms give the base on which human activities are placed. The plains are utilized for agriculture. Plateaus provide forests and minerals. Mountains provide pastures, forests, tourist spots and are sources of rivers providing water to lowlands. Climate influences our house types, clothing and food habits. The weather has a thoughtful consequence on vegetation, cropping pattern, domestic animals farming, and several industries, etc. Human beings have developed technologies which adjust climatic elements in a restricted space such as air conditioners and coolers. Temperature and rainfall make sure the compactness of forests and quality of grassland.

For, example, in India, monsoonal rainfall sets the cultivation measure in motion. Rainfall recharges the soil water aquifer which later provides water for agriculture and household use. We study oceans which are the storehouse of resources. Besides fish and other sea-food, oceans are rich in mineral resources. India has developed the technology for collecting manganese nodules from the oceanic bed. Soils are renewable resources, which influence a number of economic activities such as agriculture. The fertility of the soil is both naturally determined and culturally induced. Soils also make available the foundation for the biosphere accommodating plants, animals, and microorganisms.

So, Geography is the study of the Earth's overall surroundings and the creature that live in those environments. Physical geographers concern geomorphologic principals to study how landforms have changed in history, but ever more such principals are significant for contemporary applications.

2

Effects of Climate and Weather

Climate is the long-term weather pattern in an area, typically averaged over 30 years.

More rigorously, it is the mean and variability of meteorological variables over a time spanning from months to millions of years.

Some of the meteorological variables that are commonly measured are temperature, humidity, atmospheric pressure, wind, and precipitation. In a broader sense, climate is the state of the components of the climate system, including the atmosphere, the hydrosphere, the cryosphere, the lithosphere and the biosphere and the interactions between them.

The climate of a location is affected by its latitude/longitude, terrain, altitude, and nearby water bodies and their currents.

Climates can be classified according to the average and typical variables, most commonly temperature and precipitation. The most widely used classification scheme was the Köppen climate classification. The Thornthwaite system, in use since 1948, incorporates evapotranspiration along with temperature and precipitation information and is used in studying biological diversity and how climate change affects it. Finally, the Bergeron and Spatial Synoptic Classification systems focus on the origin of air masses that define the climate of a region.

Paleoclimatology is the study of ancient climates. Since very few direct observations of climate were available before the 19th century, paleoclimates are inferred from proxy variables. They include non-biotic evidence—such as sediments found in lake beds and ice cores—and biotic evidence—such as tree rings and coral. Climate models are mathematical models of past, present, and future climates. Climate change may occur over long and short timescales from various factors. Recent warming is discussed in global warming, which results in redistributions. For example, "a 3 °C K; 5 °F] change in mean annual temperature corresponds to a shift in isotherms of approximately 300–400 km [190–250 mi] in latitude (in the temperate zone) or 500 m [1,600 ft] in elevation. Therefore, species are expected to move upwards in elevation or towards the poles in latitude in response to shifting climate zones."

Definition

Climate is commonly defined as the weather averaged over a long period.

The standard averaging period is 30 years, but other periods may be used depending on the purpose. Climate also includes statistics other than the average, such as the magnitudes of day-to-day or year-to-year variations. The Intergovernmental Panel on Climate Change (IPCC) 2001 glossary definition is as follows:

Climate in a narrow sense is usually defined as the "average weather," or more rigorously, as the statistical description in terms of the mean and variability of relevant quantities over a period ranging from months to thousands or millions of years. The classical period is 30 years, as defined by the World Meteorological Organization (WMO). These quantities are most often surface variables such as temperature, precipitation, and wind. Climate in a wider sense is the state, including a statistical description, of the climate system.

The World Meteorological Organization (WMO) describes "climate normals" (CN) as "reference points used by climatologists to compare current climatological trends to that of the past or what is considered typical. A CN is defined as the arithmetic average of a climate element (e.g. temperature) over a 30-year period. A 30 year period is used, as it is long enough to filter out any interannual variation or anomalies, but also short enough to be able to show longer climatic trends."

The WMO originated from the International Meteorological Organization which set up a technical commission for climatology in 1929. At its 1934 Wiesbaden meeting the technical commission designated the thirty-year period from 1901 to 1930 as the reference time frame for climatological standard normals. In 1982 the WMO agreed to update climate normals, and these were subsequently completed on the basis of climate data from 1 January 1961 to 31 December 1990.

The difference between climate and weather is usefully summarized by the popular phrase "Climate is what you expect, weather is what you get."

Over historical time spans, there are a number of nearly constant variables that determine climate, including latitude, altitude, proportion of land to water, and proximity to oceans and mountains. All of these variables change only over periods of millions of years due to processes such as plate tectonics. Other climate determinants are more dynamic: the thermohaline circulation of the ocean leads to a 5 °C (9 °F) warming of the northern Atlantic Ocean compared to other ocean basins.

Other ocean currents redistribute heat between land and water on a more regional scale. The density and type of vegetation coverage affects solar heat absorption, water retention, and rainfall on a regional level. Alterations in the quantity of atmospheric greenhouse gases determines the amount of solar energy retained by the planet, leading to global warming or global cooling. The variables which determine climate are numerous and the interactions complex, but there is general agreement that the broad outlines are understood, at least insofar as the determinants of historical climate change are concerned.

Climate Classification

There are several ways to classify climates into similar regimes. Originally, climes were defined in Ancient Greece to describe the weather depending upon a location's latitude. Modern climate classification methods can be broadly divided into genetic methods, which focus on the causes of climate, and empiric methods, which focus on the effects of climate. Examples of genetic classification include methods based on the relative frequency of different air mass types or locations within synoptic weather disturbances. Examples of empiric classifications include climate zones defined by plant hardiness, evapotranspiration, or more generally the Köppen climate classification which was originally designed to identify the climates associated with certain biomes. A common shortcoming of these classification schemes is that they produce distinct boundaries between the zones they define, rather than the gradual transition of climate properties more common in nature.

Bergeron and Spatial Synoptic

The simplest classification is that involving air masses. The Bergeron classification is the most widely accepted form of air mass classification. Air mass classification involves three letters. The first letter describes its moisture properties, with c used for continental air masses (dry) and m for maritime air masses (moist). The second letter describes the thermal characteristic of its source region: T for tropical, P for polar, A for Arctic or Antarctic, M for monsoon, E for equatorial, and S for superior air (dry air formed by significant downward motion in the atmosphere). The third letter is used to designate the stability of the atmosphere. If the air mass is colder than the ground below it, it is labeled k. If the air mass is warmer than the ground below it, it is labeled w. While air mass identification was originally used in weather forecasting during the 1950s, climatologists began to establish synoptic climatologies based on this idea in 1973.

Based upon the Bergeron classification scheme is the Spatial Synoptic Classification system (SSC). There are six categories within the SSC scheme: Dry Polar (similar to continental polar), Dry Moderate (similar to maritime superior), Dry Tropical (similar to continental tropical), Moist Polar (similar to maritime polar), Moist Moderate (a hybrid between maritime polar and maritime tropical), and Moist Tropical (similar to maritime tropical, maritime monsoon, or maritime equatorial).

Köppen

The Köppen classification depends on average monthly values of temperature and precipitation. The most commonly used form of the Köppen classification has five primary types labeled A through E. These primary types are A, tropical; B, dry; C, mild mid-latitude; D, cold mid-latitude; and E, polar. The five primary classifications can be further divided into secondary classifications such as rain forest, monsoon, tropical savanna, humid subtropical, humid continental, oceanic climate, Mediterranean climate, steppe, subarctic climate, tundra, polar ice cap, and desert.

Rain forests are characterized by high rainfall, with definitions setting minimum normal annual rainfall between 1,750 millimetres (69 in) and 2,000 millimetres (79 in). Mean monthly temperatures exceed 18 °C (64 °F) during all months of the year.

A monsoon is a seasonal prevailing wind which lasts for several months, ushering in a region's rainy season. Regions within North America, South America, Sub-Saharan Africa, Australia and East Asia are monsoon regimes.

A tropical savanna is a grassland biome located in semiarid to semi-humid climate regions of subtropical and tropical latitudes, with average temperatures remain at or above 18 °C (64 °F) year round and rainfall between 750 millimetres (30 in) and 1,270 millimetres (50 in) a year. They are widespread on Africa, and are found in India, the northern parts of South America, Malaysia, and Australia.

The humid subtropical climate zone where winter rainfall (and sometimes snowfall) is associated with large storms that thewesterlies steer from west to east. Most summer rainfall occurs during thunderstorms and from occasional tropical cyclones. Humid subtropical climates lie on the east side continents, roughly between latitudes 20° and 40° degrees away from the equator.

A humid continental climate is marked by variable weather patterns and a large seasonal temperature variance. Places with more than three months of average daily temperatures above 10 °C (50 °F) and a coldest month temperature below "3 °C (27 °F)and which do not meet the criteria for an arid or semiarid climate, are classified as continental.

An oceanic climate is typically found along the west coasts at the middle latitudes of all the world's continents, and in southeastern Australia, and is accompanied by plentiful precipitation year round.

The Mediterranean climate regime resembles the climate of the lands in the Mediterranean Basin, parts of western North America, parts of Western and South Australia, in southwestern South Africa and in parts of central Chile. The climate is characterized by hot, dry summers and cool, wet winters.

A steppe is a dry grassland with an annual temperature range in the summer of up to 40 °C (104 °F) and during the winter down to"40 °C ("40 °F).

A subarctic climate has little precipitation, and monthly temperatures which are above 10 °C (50 °F) for one to three months of the year, with permafrost in large parts of the area due to the cold winters.

Winters within subarctic climates usually include up to six months of temperatures averaging below 0 °C (32 °F).

Tundra occurs in the far Northern Hemisphere, north of the taiga belt, including vast areas of northern Russia and Canada.

A polar ice cap, or polar ice sheet, is a high-latitude region of a planet or moon that is covered in ice. Ice caps form because high-latitude regions receive less energy as solar radiation from the sun than equatorial regions, resulting in lower surface temperatures.

A desert is a landscape form or region that receives very little precipitation. Deserts usually have a large diurnal and seasonal temperature range, with high or low, depending on location daytime temperatures (in summer up to 45 °C or 113 °F, and low nighttime temperatures (in winter down to 0 °C or 32 °F due to extremely low humidity. Many deserts are formed by rain shadows, as mountains block the path of moisture and precipitation to the desert.

Record

Paleoclimatology

Paleoclimatology is the study of past climate over a great period of the Earth's history. It uses evidence from ice sheets, tree rings, sediments, coral, and rocks to determine the past state of the climate. It demonstrates periods of stability and periods of change and can indicate whether changes follow patterns such as regular cycles.

Modern

Details of the modern climate record are known through the taking of measurements from such weather instruments as thermometers, barometers, and anemometers during the past few centuries. The instruments used to study weather over the modern time scale, their known error, their immediate environment, and their exposure have changed over the years, which must be considered when studying the climate of centuries past.

Climate variability

Climate variability is the term to describe variations in the mean state and other characteristics of climate (such as chances or possibility of extreme weather, etc.) "on all spatial and temporal scales beyond that of individual weather events."

Some of the variability does not appear to be caused systematically and occurs at random times. Such variability is called random variability or noise. On the other hand, periodic variability occurs relatively regularly and in distinct modes of variability or climate patterns.

There are close correlations between Earth's climate oscillations and astronomical factors (barycenter changes, solar variation, cosmic ray flux, cloud albedo feedback, Milankovic cycles), and modes of heat distribution between the ocean-atmosphere climate system. In some cases, current, historical and paleoclimatological natural oscillations may be masked by significant volcanic eruptions, impact events, irregularities in climate proxy data, positive feedback processes or anthropogenic emissions of substances such as greenhouse gases.

Over the years, the definitions of climate variability and the related term climate change have shifted. While the term climate change now implies change that is both long-term and of human causation, in the 1960s the word climate change was used for what we now describe as climate variability, that is, climatic inconsistencies and anomalies.

Thornthwaite

Devised by the American climatologist and geographer C. W. Thornthwaite, this climate classification method monitors the soil water budget using evapotranspiration. It monitors the portion of total precipitation used to nourish vegetation over a certain area. It uses indices such as a humidity index and an aridity index to determine an area's moisture regime based upon its average temperature, average rainfall, and average vegetation type. The lower the value of the index in any given area, the drier the area is.

The moisture classification includes climatic classes with descriptors such as hyperhumid, humid, subhumid, subarid, semi-arid (values of "20 to "40), and arid (values below "40). Humid regions experience more precipitation than evaporation each year, while arid regions experience greater evaporation than precipitation on an annual basis. A total of 33 percent of the Earth's landmass is considered either arid of semi-arid, including southwest North America, southwest South America, most of northern and a small part of southern Africa, southwest and portions of eastern Asia, as well as much of Australia. Studies suggest that precipitation effectiveness (PE) within the Thornthwaite moisture index is overestimated in the summer and underestimated in the winter. This index can be effectively used to determine the number of herbivore and mammal species numbers within a given area. The index is also used in studies of climate change.

Thermal classifications within the Thornthwaite scheme include microthermal, mesothermal, and megathermal regimes. A microthermal climate is one of low annual mean temperatures, generally between 0 °C (32 °F) and 14 °C (57 °F) which experiences short summers and has a potential evaporation between 14 centimetres (5.5 in) and 43 centimetres (17 in). A mesothermal climate lacks persistent heat or persistent cold, with potential evaporation between 57 centimetres (22 in) and 114 centimetres (45 in). A megathermal climate is one with persistent high temperatures and abundant rainfall, with potential annual evaporation in excess of 114 centimetres (45 in).

Paleoclimatology

Paleoclimatology is the study of past climate over a great period of the Earth's history. It uses evidence from ice sheets, tree rings, sediments, coral, and rocks to determine the past state of the climate. It demonstrates periods of stability and periods of change and can indicate whether changes follow patterns such as regular cycles.

Climate Change

Climate change is the variation in global or regional climates over time. It reflects changes in the variability or average state of the atmosphere over time scales ranging from decades to millions of years. These changes can be caused by processes internal to the Earth, external forces (e.g. variations in sunlight intensity) or, more recently, human activities.

In recent usage, especially in the context of environmental policy, the term "climate change" often refers only to changes in modern climate, including the rise in average surface temperature known as global warming. In some cases, the term is also used with a presumption of human causation, as in the United Nations Framework Convention on Climate Change (UNFCCC). The UNFCCC uses "climate variability" for non-human caused variations.

Earth has undergone periodic climate shifts in the past, including four major ice ages. These consisting of glacial periods where conditions are colder than normal, separated by interglacial periods. The accumulation of snow and ice during a glacial period increases the surface albedo, reflecting more of the Sun's energy into space and maintaining a lower atmospheric temperature. Increases in greenhouse gases, such as by volcanic activity, can increase the global temperature and produce an interglacial. Suggested causes of ice age periods include the positions of the continents, variations in the Earth's orbit, changes in the solar output, and volcanism.

Climate Models

Climate models use quantitative methods to simulate the interactions of the atmosphere, oceans, land surface and ice. They are used for a variety of purposes from study of the dynamics of the weather and climate system to projections of future climate. All climate models balance, or very nearly balance, incoming energy as short wave (including visible) electromagnetic radiation to the earth with outgoing energy as long wave (infrared) electromagnetic radiation from the earth. Any imbalance results in a change in the average temperature of the earth.

The most talked-about applications of these models in recent years have been their use to infer the consequences of increasing greenhouse gases in the atmosphere, primarily carbon dioxide. These models predict an upward trend in the global mean surface temperature, with the most rapid increase in temperature being projected for the higher latitudes of the Northern Hemisphere.

Models can range from relatively simple to quite complex:

- Simple radiant heat transfer model that treats the earth as a single point and averages outgoing energy
- this can be expanded vertically (radiative-convective models), or horizontally
- finally, (coupled) atmosphere–ocean–sea ice global climate models discretise and solve the full equations for mass and energy transfer and radiant exchange.

Climate forecasting is a way by some scientists are using to predict climate change. In 1997 the prediction division of the International Research Institute for Climate and Society at Columbia University began generating seasonal climate forecasts on a real-time basis. To produce these forecasts an extensive suite of forecasting tools was developed, including a multimodel ensemble approach that required thorough validation of each model's accuracy level in simulating interannual climate variability.

Greenhouse Effect

The greenhouse effect is a process by which thermal radiation from a planetary surface is absorbed by atmospheric greenhouse gases, and is re-radiated in all directions. Since part of this re-radiation is back towards the surface, energy is transferred to the surface and the lower atmosphere. As a result, the average surface temperature is higher than it would be if direct heating by solar radiation were the only warming mechanism.

Solar radiation at the high frequencies of visible light passes through the atmosphere to warm the planetary surface, which then emits this energy at the lower frequencies of infrared thermal radiation. Infrared radiation is absorbed by greenhouse gases, which in turn re-radiate much of the energy to the surface and lower atmosphere.

The mechanism is named after the effect of solar radiation passing through glass and warming a greenhouse, but the way it retains heat is fundamentally different as a greenhouse works by reducing airflow, isolating the warm air inside the structure so that heat is not lost by convection.

The existence of the greenhouse effect was argued for by Joseph Fourier in 1824. The argument and the evidence was further strengthened by Claude Pouillet in 1827 and 1838, and definitively proved experimentally by John Tyndall in 1859, and more fully quantified by Svante Arrheniusin 1896.

If an ideal thermally conductive blackbody was the same distance from the Sun as the Earth is, it would have a temperature of about 5.3 °C. However, since the Earth reflects about 30% (or 28%) of the incoming sunlight, the planet's effective temperature (the temperature of a blackbody that would emit the same amount of radiation) is about "18 or "19 °C, about 33°C below the actual surface temperature of about 14 °C or 15 °C. The mechanism that produces this difference between the actual surface temperature and the effective temperature is due to the atmosphere and is known as the greenhouse effect.

Earth's natural greenhouse effect makes life as we know it possible. However, human activities, primarily the burning of fossil fuels and clearing of forests, have greatly intensified the natural greenhouse effect, causing global warming.

Basic Mechanism

The Earth receives energy from the Sun in the form UV, visible, and near IR radiation, most of which passes through the atmosphere without being absorbed. Of the total amount of energy available at the top of the atmosphere (TOA), about 50% is absorbed at the Earth's surface. Because it is warm, the surface radiates far IR thermal radiation that consists of wavelengths that are predominantly much longer than the wavelengths that were absorbed. Most of this thermal radiation is absorbed by the atmosphere and re-radiated both upwards and downwards; that radiated downwards is absorbed by the Earth's surface. This trapping of long-wavelength thermal radiation leads to a higher equilibrium temperature than if the atmosphere were absent.

This highly simplified picture of the basic mechanism needs to be qualified in a number of ways, none of which affect the fundamental process.

- The incoming radiation from the Sun is mostly in the form of visible light and nearby wavelengths, largely in the range 0.2–4 ìm, corresponding to the Sun's radiative temperature of 6,000 K. Almost half the radiation is in the form of "visible" light, which our eyes are adapted to use.
- About 50% of the Sun's energy is absorbed at the Earth's surface and the rest is reflected or absorbed by the atmosphere. The reflection of light back into space—largely by clouds—does not much affect the basic mechanism; this light, effectively, is lost to the system.
- The absorbed energy warms the surface. Simple presentations of the greenhouse effect, such as the idealized greenhouse model, show this heat being lost as thermal radiation. The reality is more complex: the atmosphere near the surface is largely opaque to thermal radiation (with important exceptions for "window"

bands), and most heat loss from the surface is by sensible heat and latent heat transport. Radiative energy losses become increasingly important higher in the atmosphere largely because of the decreasing concentration of water vapor, an important greenhouse gas. It is more realistic to think of the greenhouse effect as applying to a "surface" in the mid-troposphere, which is effectively coupled to the surface by a lapse rate.

- Within the region where radiative effects are important the description given by the idealized greenhouse model becomes realistic: The surface of the Earth, warmed to a temperature around 255 K, radiates long-wavelength, infrared heat in the range 4–100 ìm. At these wavelengths, greenhouse gases that were largely transparent to incoming solar radiation are more absorbent. Each layer of atmosphere with greenhouses gases absorbs some of the heat being radiated upwards from lower layers. To maintain its own equilibrium, it re-radiates the absorbed heat in all directions, both upwards and downwards. This results in more warmth below, while still radiating enough heat back out into deep space from the upper layers to maintain overall thermal equilibrium. Increasing the concentration of the gases increases the amount of absorption and re-radiation, and thereby further warms the layers and ultimately the surface below.
- Greenhouse gases—including most diatomic gases with two different atoms (such as carbon monoxide, CO) and all gases with three or more atoms—are able to absorb and emit infrared radiation. Though more than 99% of the dry atmosphere is IR transparent (because the main constituents—N2, O2, and Ar—are not able to directly absorb or emit infrared radiation), intermolecular collisions cause the energy absorbed and emitted by the greenhouse gases to be shared with the other, non-IR-active, gases.
- The simple picture assumes equilibrium. In the real world there is the diurnal cycle as well as seasonal cycles and weather. Solar heating only applies during daytime. During the night, the atmosphere cools somewhat, but not greatly, because its emissivity is low, and during the day the atmosphere warms. Diurnal temperature changes decrease with height in the atmosphere.

Greenhouse Gases

By their percentage contribution to the greenhouse effect on Earth the four major gases are:

- water vapor, 36–70%
- carbon dioxide, 9–26%
- methane, 4–9%
- ozone, 3–7%

The major non-gas contributor to the Earth's greenhouse effect, clouds, also absorb and emit infrared radiation and thus have an effect on radiative properties of the atmosphere.

Role in climate change

Strengthening of the greenhouse effect through human activities is known as the enhanced (or anthropogenic) greenhouse effect. This increase in radiative forcing from human activity is attributable mainly to increased atmospheric carbon dioxide levels.

CO2 is produced by fossil fuel burning and other activities such as cement production and tropical deforestation. Measurements of CO2 from the Mauna Loa observatory show that concentrations have increased from about 313 ppm in 1960 to about 389 ppm in 2010. The current observed amount of CO2 exceeds the geological record maxima (~300 ppm) from ice core data. The effect of combustion-produced carbon dioxide on the global climate, a special case of the greenhouse effect first described in 1896 by Svante Arrhenius, has also been called theCallendar effect.

Because it is a greenhouse gas, elevated CO2 levels contribute to additional absorption and emission of thermal infrared in the atmosphere, which produce net warming. According to the latest Assessment Report from the Intergovernmental Panel on Climate Change, "most of the observed increase in globally averaged temperatures since the mid-20th century is very likely due to the observed increase in anthropogenic greenhouse gas concentrations".

Over the past 800,000 years, ice core data shows that carbon dioxide has varied from values as low as 180 parts per million (ppm) to the pre-industrial level of 270ppm. Paleoclimatologists consider variations in carbon dioxide concentration to be a fundamental factor influencing climate variations over this time scale.

Real Greenhouses

The "greenhouse effect" is named by analogy to greenhouses. The greenhouse effect and a real greenhouse are similar in that they both limit the rate of thermal energy flowing out of the system, but the mechanisms by which heat is retained are different. A greenhouse works primarily by preventing absorbed heat from leaving the structure through convection, i.e. sensible heat transport. The greenhouse effect heats the earth because greenhouse gases absorb outgoing radiative energy and re-emit some of it back towards earth.

A greenhouse is built of any material that passes sunlight, usually glass, or plastic. It mainly heats up because the Sun warms the ground inside, which then warms the air in the greenhouse. The air continues to heat because it is confined within the greenhouse, unlike the environment outside the greenhouse where warm air near the surface rises and mixes with cooler air aloft. This can be demonstrated by opening a small window near the roof of a greenhouse: the temperature will drop considerably. It has also been demonstrated experimentally (R. W. Wood, 1909) that a "greenhouse" with a cover of rock salt (which is transparent to infra red) heats up an enclosure similarly to one with a glass cover. Thus greenhouses work primarily by preventing convective cooling.

In the greenhouse effect, rather than retaining (sensible) heat by physically preventing movement of the air, greenhouse gases act to warm the Earth by re-radiating some of the energy back towards the surface. This process may exist in real greenhouses, but is comparatively unimportant there.

Bodies Other than Earth

In our solar system, Mars, Venus, and the moon Titan also exhibit greenhouse effects. Titan has an anti-greenhouse effect, in that its atmosphere absorbs solar radiation but is relatively transparent to infrared radiation. Pluto also exhibits behavior superficially similar to the anti-greenhouse effect.

A runaway greenhouse effect occurs if positive feedbacks lead to the evaporation of all greenhouse gases into the atmosphere. A runaway greenhouse effect involving carbon dioxide and water vapor is thought to have occurred on Venus.

Weather

Weather is the state of the atmosphere, to the degree that it is hot or cold, wet or dry, calm or stormy, clear or cloudy. Most weather phenomena occur in the troposphere, just below the stratosphere. Weather refers, generally, to day-to-day temperature and precipitation activity, whereas climate is the term for the average atmospheric conditions over longer periods of time. When used without qualification, "weather" is understood to be the weather of Earth.

Weather is driven by density (temperature and moisture) differences between one place and another. These differences can occur due to the sun angle at any particular spot, which varies by latitude from the tropics. The strong temperature contrast between polar and tropical air gives rise to the jet stream. Weather systems in the mid-latitudes, such as extratropical cyclones, are caused by instabilities of the jet stream flow. Because the Earth's axis is tilted relative to its orbital plane, sunlight is incident at different angles at different times of the year. On Earth's surface, temperatures usually range ±40 °C (100 °F to "40 °F) annually. Over thousands of years, changes in Earth'sorbit affect the amount and distribution of solar energy received by the Earth and influence long-term climate and global climate change.

Surface temperature differences in turn cause pressure differences. Higher altitudes are cooler than lower altitudes due to differences in compressional heating. Weather forecasting is the application of science and technology to predict the state of the atmosphere for a future time and a given location. The atmosphere is a chaotic system, so small changes to one part of the system can grow to have large effects on the system as a whole. Human attempts to control the weather have occurred throughout human history, and there is evidence that human activity such as agriculture and industry has inadvertently modified weather patterns.

Studying how the weather works on other planets has been helpful in understanding how weather works on Earth. A famous landmark in the Solar System, Jupiter's Great Red Spot, is an anticyclonic storm known to have existed for at least 300 years. However, weather is not limited to planetary bodies. A star's corona is constantly being lost to space, creating what is essentially a very thin atmosphere throughout the Solar System. The movement of mass ejected from the Sun is known as the solar wind.

On Earth, common weather phenomena include wind, cloud, rain, snow, fog and dust storms. Less common events include natural disasters such as tornadoes, hurricanes, typhoons and ice storms. Almost all familiar weather phenomena occur in the troposphere (the lower part of the atmosphere). Weather does occur in the stratosphere and can affect weather lower down in the troposphere, but the exact mechanisms are poorly understood.

Weather occurs primarily due to density (temperature and moisture) differences between one place to another. These differences can occur due to the sun angle at any particular spot, which varies by latitude from the tropics. In other words, the farther from the tropics you lie, the lower the sun angle is, which causes those locations to be cooler due to the indirect sunlight. The strong temperature contrast between polar and tropical air gives rise to the jet stream. Weather systems in the mid-latitudes, such as extratropical cyclones, are caused by instabilities of the jet stream flow. Weather systems in the tropics, such as monsoons or organized thunderstorm systems, are caused by different processes.

Because the Earth's axis is tilted relative to its orbital plane, sunlight is incident at different angles at different times of the year. In June the Northern Hemisphere is tilted towards thesun, so at any given Northern Hemisphere latitude sunlight falls more directly on that spot than in December. This effect causes seasons. Over thousands to hundreds of thousands of years, changes in Earth's orbital parameters affect the amount and distribution of solar energy received by the Earth and influence long-term climate.

The uneven solar heating (the formation of zones of temperature and moisture gradients, or frontogenesis) can also be due to the weather itself in the form of cloudiness and precipitation. Higher altitudes are cooler than lower altitudes, which is explained by the lapse rate. On local scales, temperature differences can occur because different surfaces (such as oceans, forests, ice sheets, or man-made objects) have differing physical characteristics such as reflectivity, roughness, or moisture content.

Surface temperature differences in turn cause pressure differences. A hot surface heats the air above it and the air expands, lowering the air pressure and its density. The resulting horizontal pressure gradient accelerates the air from high to low pressure, creating wind, and Earth's rotation then causes curvature of the flow via the Coriolis effect. The simple systems thus formed can then display emergent behaviour to produce more complex systems and thus other weather phenomena. Large scale examples include the Hadley cell while a smaller scale example would be coastal breezes.

The atmosphere is a chaotic system, so small changes to one part of the system can grow to have large effects on the system as a whole. This makes it difficult to accurately predict weather more than a few days in advance, though weather forecasters are continually working to extend this limit through the scientific study of weather, meteorology. It is theoretically impossible to make useful day-to-day predictions more than about two weeks ahead, imposing an upper limit to potential for improved prediction skill.

Shaping the Planet Earth

Weather is one of the fundamental processes that shape the Earth. The process of weathering breaks down the rocks and soils into smaller fragments and then into their constituent substances. These are then free to take part in chemical reactions that can affect the surface further (such as acid rain) or are reformed into other rocks and soils. In this way, weather plays a major role in erosion of the surface.

Effects on Populations

Weather has played a large and sometimes direct part in human history. Aside from climatic changes that have caused the gradual drift of populations (for example the desertification of the Middle East, and the formation of land bridges during glacial periods), extreme weather events have caused smaller scale population movements and intruded directly in historical events. One such event is the saving of Japan from invasion by the Mongol fleet of Kublai Khan by the Kamikaze winds in 1281.

French claims to Florida came to an end in 1565 when a hurricane destroyed the French fleet, allowing Spain to conquer Fort Caroline. More recently, Hurricane Katrinaredistributed over one million people from the central Gulf coast elsewhere across the United States, becoming the largest diaspora in the history of the United States.

The Little Ice Age caused crop failures and famines in Europe. The 1690s saw the worst famine in France since the Middle Ages. Finland suffered a severe famine in 1696–1697, during which about one-third of the Finnish population died.

Effects on Individuals

The human body is negatively affected by extremes in temperature, humidity, and wind.

Forecasting

Weather forecasting is the application of science and technology to predict the state of the atmosphere for a future time and a given location. Human beings have attempted to predict the weather informally for millennia, and formally since at least the nineteenth century. Weather forecasts are made by collecting quantitative data about the current state of the atmosphere and using scientific understanding of atmospheric processes to project how the atmosphere will evolve.

Once an all-human endeavor based mainly upon changes in barometric pressure, current weather conditions, and sky condition, forecast models are now used to determine future conditions. Human input is still required to pick the best possible forecast model to base the forecast upon, which involves pattern recognition skills, teleconnections, knowledge of model performance, and knowledge of model biases.

The chaotic nature of the atmosphere, the massive computational power required to solve the equations that describe the atmosphere, error involved in measuring the initial conditions, and an incomplete understanding of atmospheric processes mean that forecasts become less accurate as the difference in current time and the time for which the forecast is being made (the range of the forecast) increases. The use of ensembles and model consensus helps to narrow the error and pick the most likely outcome.

There are a variety of end users to weather forecasts. Weather warnings are important forecasts because they are used to protect life and property. Forecasts based on temperature and precipitation are important to agriculture, and therefore to commodity traders within stock markets. Temperature forecasts are used by utility companies to estimate demand over coming days. On an everyday basis, people use weather forecasts to determine what to wear on a given day. Since outdoor activities are severely curtailed by heavy rain, snow and the wind chill, forecasts can be used to plan activities around these events, and to plan ahead and survive them.

Modification

The aspiration to control the weather is evident throughout human history: from ancient rituals intended to bring rain for crops to the U.S. Military Operation Popeye, an attempt to disruptsupply lines by lengthening the North Vietnamese monsoon. The most successful attempts at influencing weather involve cloud seeding; they include the fog- and low stratus dispersion techniques employed by major airports, techniques used to increase winter precipitation over mountains, and techniques to suppress hail.

A recent example of weather control was China's preparation for the 2008 Summer Olympic Games. China shot 1,104 rain dispersal rockets from 21 sites in the city of Beijing in an effort to keep rain away from the opening ceremony of the games on 8 August 2008. Guo Hu, head of the Beijing Municipal Meteorological Bureau (BMB), confirmed the success of the operation with 100 millimeters falling inBaoding City of Hebei Province, to the southwest and Beijing's Fangshan District recording a rainfall of 25 millimeters.

Whereas there is inconclusive evidence for these techniques' efficacy, there is extensive evidence that human activity such as agriculture and industry results in inadvertent weather modification:

- Acid rain, caused by industrial emission of sulfur dioxide and nitrogen oxides into the atmosphere, adversely affects freshwater lakes, vegetation, and structures.
- Anthropogenic pollutants reduce air quality and visibility.
- Climate change caused by human activities that emit greenhouse gases into the air is expected to affect the frequency of extreme weather events such as drought, extreme temperatures, flooding, high winds, and severe storms.

The effects of inadvertent weather modification may pose serious threats to many aspects of civilization, including ecosystems, natural resources, food and fiber production, economic development, and human health.

Extremes on Earth

On Earth, temperatures usually range ±40 °C (100 °F to "40 °F) annually. The range of climates and latitudes across the planet can offer extremes of temperature outside this range. The coldest air temperature ever recorded on Earth is "89.2 °C ("128.6 °F), at Vostok Station, Antarctica on 21 July 1983.

The hottest air temperature ever recorded was 57.7 °C (135.9 °F) at 'Aziziya, Libya, on 13 September 1922, but that reading is queried. The highest recorded average annual temperature was 34.4 °C (93.9 °F) at Dallol, Ethiopia. The coldest recorded average annual temperature was "55.1 °C ("67.2 °F) at Vostok Station, Antarctica. The coldest average annual temperature in a permanently inhabited location is at Eureka, Nunavut, in Canada, where the annual average temperature is "19.7 °C("3.5 °F).

Extra Terrestrial within the Solar System

Studying how the weather works on other planets has been seen as helpful in understanding how it works on Earth. Weather on other planets follows many of the same physical principles as weather on Earth, but occurs on different scales and in atmospheres having different chemical composition. The Cassini–Huygens mission to Titan discovered clouds formed from methane or ethane which deposit rain composed of liquid methane and other organic compounds. Earth's atmosphere includes six latitudinal circulation zones, three in each hemisphere. In contrast, Jupiter's banded appearance shows many such zones, Titan has a single jet stream near the 50th parallel north latitude, and Venus has a single jet near the equator.

One of the most famous landmarks in the Solar System, Jupiter's Great Red Spot, is an anticyclonic storm known to have existed for at least 300 years. On other gas giants, the lack of a surface allows the wind to reach enormous speeds: gusts of up to 600 metres per second (about 2,100 km/h or 1,300 mph) have been measured on the planet Neptune.

This has created a puzzle for planetary scientists. The weather is ultimately created by solar energy and the amount of energy received by Neptune is only about 1D 900 of that received by Earth, yet the intensity of weather phenomena on Neptune is far greater than on Earth. The strongest planetary winds discovered so far are on the extrasolar planet HD 189733 b, which is thought to have easterly winds moving at more than 9,600 kilometres per hour (6,000 mph).

Space Weather

Weather is not limited to planetary bodies. Like all stars, the sun's corona is constantly being lost to space, creating what is essentially a very thin atmosphere throughout the Solar System. The movement of mass ejected from the Sun is known as the solar wind. Inconsistencies in this wind and larger events on the surface of the star, such as coronal mass ejections, form a system that has features analogous to conventional weather systems (such as pressure and wind) and is generally known as space weather.

Coronal mass ejections have been tracked as far out in the solar system as Saturn. The activity of this system can affect planetary atmospheres and occasionally surfaces. The interaction of the solar wind with the terrestrial atmosphere can produce spectacular aurorae, and can play havoc with electrically sensitive systems such as electricity grids and radio signals.

Weather forecasting

Weather forecasting is the application of science and technology to predict the conditions of the atmosphere for a given location and time. People have attempted to predict the weather informally for millennia and formally since the 19th century. Weather forecasts are made by collecting quantitative data about the current state of the atmosphere, land, and ocean and using meteorology to project how the atmosphere will change at a given place.

Once calculated manually based mainly upon changes in barometric pressure, current weather conditions, and sky condition or cloud cover, weather forecasting now relies on computer-based models that take many atmospheric factors into account.

Human input is still required to pick the best possible forecast model to base the forecast upon, which involves pattern recognition skills, teleconnections, knowledge of model performance, and knowledge of model biases. The inaccuracy of forecasting is due to the chaotic nature of the atmosphere, the massive computational power required to solve the equations that describe the atmosphere, the land, and the ocean, the error involved in measuring the initial conditions, and an incomplete understanding of atmospheric and related processes. Hence, forecasts become less accurate as the difference between current time and the time for which the forecast is being made (the range of the forecast) increases. The use of ensembles and model consensus help narrow the error and provide confidence level in the forecast.

There is a vast variety of end uses to weather forecasts. Weather warnings are important forecasts because they are used to protect life and property. Forecasts based on temperature and precipitation are important to agriculture, and therefore to traders within commodity markets. Temperature forecasts are used by utility companies to estimate demand over coming days. On an everyday basis, many use weather forecasts to determine what to wear on a given day. Since outdoor activities are severely curtailed by heavy rain, snow and wind chill, forecasts can be used to plan activities around these events, and to plan ahead and survive them.

Weather forecasting is a part of the economy, for example, in 2009, the US spent approximately $5.1 billion on weather forecasting, producing benefits estimated at six times as much.

Ancient forecasting

For millennia, people have tried to forecast the weather. In 650 BC, the Babylonians predicted the weather from cloud patterns as well as astrology. In about 350 BC, Aristotle described weather patterns in Meteorologica.

Later, Theophrastus compiled a book on weather forecasting, called the Book of Signs.

Chinese weather prediction lore extends at least as far back as 300 BC,

which was also around the same time ancient Indian astronomers developed weather-prediction methods.

In New Testament times, Jesus himself referred to deciphering and understanding local weather patterns, by saying, "When evening comes, you say, 'It will be fair weather, for the sky is red', and in the morning, 'Today it will be stormy, for the sky is red and overcast.' You know how to interpret the appearance of the sky, but you cannot interpret the signs of the times."

In 904 AD, Ibn Wahshiyya's Nabatean Agriculture, translated into Arabic from an earlier Aramaic work, discussed the weather forecasting of atmospheric changes and signs from the planetary astral alterations; signs of rain based on observation of the lunar phases; and weather forecasts based on the movement of winds.

Ancient weather forecasting methods usually relied on observed patterns of events, also termed pattern recognition. For example, it was observed that if the sunset was particularly red, the following day often brought fair weather. This experience accumulated over the generations to produce weather lore. However, not all[which?] of these predictions prove reliable, and many of them have since been found not to stand up to rigorous statistical testing.

Modern methods

It was not until the invention of the electric telegraph in 1835 that the modern age of weather forecasting began.

Before that, the fastest that distant weather reports could travel was around 160 kilometres per day (100 mi/d), but was more typically 60–120 kilometres per day (40–75 mi/day) (whether by land or by sea).

By the late 1840s, the telegraph allowed reports of weather conditions from a wide area to be received almost instantaneously, allowing forecasts to be made from knowledge of weather conditions further upwind.

The two men credited with the birth of forecasting as a science were an officer of the Royal Navy Francis Beaufort and his protégé Robert FitzRoy. Both were influential men in British naval and governmental circles, and though ridiculed in the press at the time, their work gained scientific credence, was accepted by the Royal Navy, and formed the basis for all of today's weather forecasting knowledge.

Beaufort developed the Wind Force Scale and Weather Notation coding, which he was to use in his journals for the remainder of his life. He also promoted the development of reliable tide tables around British shores, and with his friend William Whewell, expanded weather record-keeping at 200 British coast guard stations.

Robert FitzRoy was appointed in 1854 as chief of a new department within the Board of Trade to deal with the collection of weather data at sea as a service to mariners. This was the forerunner of the modern Meteorological Office.

All ship captains were tasked with collating data on the weather and computing it, with the use of tested instruments that were loaned for this purpose.

A storm in October 1859 that caused the loss of the Royal Charter inspired FitzRoy to develop charts to allow predictions to be made, which he called "forecasting the weather", thus coining the term "weather forecast".

Fifteen land stations were established to use the telegraph to transmit to him daily reports of weather at set times leading to the first gale warning service. His warning service for shipping was initiated in February 1861, with the use of telegraph communications. The first daily weather forecasts were published in The Times in 1861.

In the following year a system was introduced of hoisting storm warning cones at the principal ports when a gale was expected.

The "Weather Book" which FitzRoy published in 1863 was far in advance of the scientific opinion of the time.

As the electric telegraph network expanded, allowing for the more rapid dissemination of warnings, a national observational network was developed, which could then be used to provide synoptic analyses. Instruments to continuously record variations in meteorological parameters using photography were supplied to the observing stations from Kew Observatory – these cameras had been invented by Francis Ronalds in 1845 and his barograph had earlier been used by FitzRoy.

To convey accurate information, it soon became necessary to have a standard vocabulary describing clouds; this was achieved by means of a series of classifications first achieved by Luke Howard in 1802, and standardized in the International Cloud Atlas of 1896.

Numerical prediction

It was not until the 20th century that advances in the understanding of atmospheric physics led to the foundation of modern numerical weather prediction. In 1922, English scientist Lewis Fry Richardson published "Weather Prediction By Numerical Process", after finding notes and derivations he worked on as an ambulance driver in World War I. He described therein how small terms in the prognostic fluid dynamics equations governing atmospheric flow could be neglected, and a finite differencing scheme in time and space could be devised, to allow numerical prediction solutions to be found.

Richardson envisioned a large auditorium of thousands of people performing the calculations and passing them to others. However, the sheer number of calculations required was too large to be completed without the use of computers, and the size of the grid and time steps led to unrealistic results in deepening systems. It was later found, through numerical analysis, that this was due to numerical instability.

The first computerised weather forecast was performed by a team composed of American meteorologists Jule Charney, Philip Thompson, Larry Gates, and Norwegian meteorologist Ragnar Fjørtoft, applied mathematician John von Neumann, and ENIAC programmer Klara Dan von Neumann.

Practical use of numerical weather prediction began in 1955, spurred by the development of programmable electronic computers.

Broadcasts

The first ever daily weather forecasts were published in The Times on August 1, 1861, and the first weather maps were produced later in the same year.

In 1911, the Met Office began issuing the first marine weather forecasts via radio transmission. These included gale and storm warnings for areas around Great Britain.

In the United States, the first public radio forecasts were made in 1925 by Edward B. "E.B." Rideout, on WEEI, the Edison Electric Illuminating station in Boston.

Rideout came from the U.S. Weather Bureau, as did WBZ weather forecaster G. Harold Noyes in 1931.

The world's first televised weather forecasts, including the use of weather maps, were experimentally broadcast by the BBC in November 1936.

This was brought into practice in 1949, after World War II.

George Cowling gave the first weather forecast while being televised in front of the map in 1954.

In America, experimental television forecasts were made by James C. Fidler in Cincinnati in either 1940 or 1947 on the DuMont Television Network.

In the late 1970s and early 80s, John Coleman, the first weatherman on ABC-TV's Good Morning America, pioneered the use of on-screen weather satellite information and computer graphics for television forecasts.

Coleman was a co-founder of The Weather Channel (TWC) in 1982. TWC is now a 24-hour cable network. Some weather channels have started broadcasting on live broadcasting programs such as YouTube and Periscope to reach more viewers.

How models create forecasts

The basic idea of numerical weather prediction is to sample the state of the fluid at a given time and use the equations of fluid dynamics and thermodynamics to estimate the state of the fluid at some time in the future. The main inputs from country-based weather services are surface observations from automated weather stations at ground level over land and from weather buoys at sea. The World Meteorological Organization acts to standardize the instrumentation, observing practices and timing of these observations worldwide. Stations either report hourly in METAR reports, or every six hours in SYNOP reports.

Sites launch radiosondes, which rise through the depth of the troposphere and well into the stratosphere.

Data from weather satellites are used in areas where traditional data sources are not available.

Compared with similar data from radiosondes, the satellite data has the advantage of global coverage, however at a lower accuracy and resolution.

Meteorological radar provide information on precipitation location and intensity, which can be used to estimate precipitation accumulations over time.

Additionally, if a pulse Doppler weather radar is used then wind speed and direction can be determined.

Commerce provides pilot reports along aircraft routes, and ship reports along shipping routes. Research flights using reconnaissance aircraft fly in and around weather systems of interest such as tropical cyclones. Reconnaissance aircraft are also flown over the open oceans during the cold season into systems that cause significant uncertainty in forecast guidance, or are expected to be of high impact 3–7 days into the future over the downstream continent.

Models are initialized using this observed data. The irregularly spaced observations are processed by data assimilation and objective analysis methods, which perform quality control and obtain values at locations usable by the model's mathematical algorithms (usually an evenly spaced grid). The data are then used in the model as the starting point for a forecast. Commonly, the set of equations used to predict the physics and dynamics of the atmosphere are called primitive equations. These equations are initialized from the analysis data and rates of change are determined. The rates of change predict the state of the atmosphere a short time into the future. The equations are then applied to this new atmospheric state to find new rates of change, and these new rates of change predict the atmosphere at a yet further time into the future. This time stepping procedure is continually repeated until the solution reaches the desired forecast time.

The length of the time step chosen within the model is related to the distance between the points on the computational grid, and is chosen to maintain numerical stability. Time steps for global models are on the order of tens of minutes, while time steps for regional models are between one and four minutes. The global models are run at varying times into the future. The Met Office's Unified Model is run six days into the future, the European Centre for Medium-Range Weather Forecasts model is run out to 10 days into the future, while the Global Forecast System model run by the Environmental Modeling Center is run 16 days into the future. The visual output produced by a model solution is known as a prognostic chart, or prog. The raw output is often modified before being presented as the forecast. This can be in the form of statistical techniques to remove known biases in the model, or of adjustment to take into account consensus among other numerical weather forecasts. MOS or model output statistics is a technique used to interpret numerical model output and produce site-specific guidance. This guidance is presented in coded numerical form, and can be obtained for nearly all National Weather Service reporting stations in the United States. As proposed by Edward Lorenz in 1963, long range forecasts, those made at a range of two weeks or more, are impossible to definitively predict the state of the atmosphere, owing to the chaotic nature of the fluid dynamics equations involved. In numerical models, extremely small errors in initial values double roughly every five days for variables such as temperature and wind velocity.

Essentially, a model is a computer program that produces meteorological information for future times at given locations and altitudes. Within any modern model is a set of equations, known as the primitive equations, used to predict the future state of the atmosphere. These equations—along with the ideal gas law—are used to evolve the density, pressure, and potential temperature scalar fields and the velocity vector field of the atmosphere through time. Additional transport equations for pollutants and other aerosols are included in some primitive-equation mesoscale models as well. The equations used are nonlinear partial differential equations, which are impossible to solve exactly through analytical methods, with the exception of a few idealized cases. Therefore, numerical methods obtain approximate solutions. Different models use different solution methods: some global models use spectral methods for the horizontal dimensions and finite difference methods for the vertical dimension, while regional models and other global models usually use finite-difference methods in all three dimensions.

Techniques

Persistence

The simplest method of forecasting the weather, persistence, relies upon today's conditions to forecast the conditions tomorrow. This can be a valid way of forecasting the weather when it is in a steady state, such as during the summer season in the tropics. This method of forecasting strongly depends upon the presence of a stagnant weather pattern. Therefore, when in a fluctuating weather pattern, this method of forecasting becomes inaccurate. It can be useful in both short range forecasts and long range forecasts.

Use of a barometer

Measurements of barometric pressure and the pressure tendency (the change of pressure over time) have been used in forecasting since the late 19th century. The larger the change in pressure, especially if more than 3.5 hPa (2.6 mmHg), the larger the change in weather can be expected. If the pressure drop is rapid, a low pressure system is approaching, and there is a greater chance of rain. Rapid pressure rises are associated with improving weather conditions, such as clearing skies.

Looking at the sky

Along with pressure tendency, the condition of the sky is one of the more important parameters used to forecast weather in mountainous areas. Thickening of cloud cover or the invasion of a higher cloud deck is indicative of rain in the near future. High thin cirrostratus clouds can create halos around the sun or moon, which indicates an approach of a warm front and its associated rain. Morning fog portends fair conditions, as rainy conditions are preceded by wind or clouds that prevent fog formation. The approach of a line of thunderstorms could indicate the approach of a cold front. Cloud-free skies are indicative of fair weather for the near future. A bar can indicate a coming tropical cyclone. The use of sky cover in weather prediction has led to various weather lore over the centuries.

Nowcasting

The forecasting of the weather within the next six hours is often referred to as nowcasting. In this time range it is possible to forecast smaller features such as individual showers and thunderstorms with reasonable accuracy, as well as other features too small to be resolved by a computer model. A human given the latest radar, satellite and observational data will be able to make a better analysis of the small scale features present and so will be able to make a more accurate forecast for the following few hours. However, there are now expert systems using those data and mesoscale numerical model to make better extrapolation, including evolution of those features in time. Accuweather is known for a Minute-Cast, which is a minute-by-minute precipitation forecast for the next two hours.

Use of forecast models

In the past, the human forecaster was responsible for generating the entire weather forecast based upon available observations. Today, human input is generally confined to choosing a model based on various parameters, such as model biases and performance. Using a consensus of forecast models, as well as ensemble members of the various models, can help reduce forecast error. However, regardless how small the average error becomes with any individual system, large errors within any particular piece of guidance are still possible on any given model run. Humans are required to interpret the model data into weather forecasts that are understandable to the end user. Humans can use knowledge of local effects that may be too small in size to be resolved by the model to add information to the forecast. While increasing accuracy of forecast models implies that humans may no longer be needed in the forecast process at some point in the future, there is currently still a need for human intervention.

Analog technique

The analog technique is a complex way of making a forecast, requiring the forecaster to remember a previous weather event that is expected to be mimicked by an upcoming event. What makes it a difficult technique to use is that there is rarely a perfect analog for an event in the future. Some call this type of forecasting pattern recognition. It remains a useful method of observing rainfall over data voids such as oceans, as well as the forecasting of precipitation amounts and distribution in the future. A similar technique is used in medium range forecasting, which is known as teleconnections, when systems in other locations are used to help pin down the location of another system within the surrounding regime. An example of teleconnections are by using El Niño-Southern Oscillation (ENSO) related phenomena.

Communicating forecasts to the public

Most end users of forecasts are members of the general public. Thunderstorms can create strong winds and dangerous lightning strikes that can lead to deaths, power outages, and widespread hail damage. Heavy snow or rain can bring transportation and commerce to a stand-still, as well as cause flooding in low-lying areas. Excessive heat or cold waves can sicken or kill those with inadequate utilities, and droughts can impact water usage and destroy vegetation.

Several countries employ government agencies to provide forecasts and watches/ warnings/advisories to the public to protect life and property and maintain commercial interests. Knowledge of what the end user needs from a weather forecast must be taken into account to present the information in a useful and understandable way. Examples include the National Oceanic and Atmospheric Administration's

National Weather Service (NWS) and Environment Canada's Meteorological Service (MSC). Traditionally, newspaper, television, and radio have been the primary outlets for presenting weather forecast information to the public. In addition, some cities had weather beacons. Increasingly, the internet is being used due to the vast amount of specific information that can be found. In all cases, these outlets update their forecasts on a regular basis.

Severe weather alerts and advisories

A major part of modern weather forecasting is the severe weather alerts and advisories that the national weather services issue in the case that severe or hazardous weather is expected. This is done to protect life and property. Some of the most commonly known of severe weather advisories are the severe thunderstorm and tornado warning, as well as the severe thunderstorm and tornado watch. Other forms of these advisories include winter weather, high wind, flood, tropical cyclone, and fog. Severe weather advisories and alerts are broadcast through the media, including radio, using emergency systems as the Emergency Alert System, which break into regular programming.

Low temperature forecast

The low temperature forecast for the current day is calculated using the lowest temperature found between 7 pm that evening through 7 am the following morning. So, in short, today's forecasted low is most likely tomorrow's low temperature.

Specialist forecasting

There are a number of sectors with their own specific needs for weather forecasts and specialist services are provided to these users as given below:

Air traffic

Because the aviation industry is especially sensitive to the weather, accurate weather forecasting is essential. Fog or exceptionally low ceilings can prevent many aircraft from landing and taking off. Turbulence and icing are also significant in-flight hazards. Thunderstorms are a problem for all aircraft because of severe turbulence due to their updrafts and outflow boundaries, icing due to the heavy precipitation, as well as large hail, strong winds, and lightning, all of which can cause severe damage to an aircraft in flight. Volcanic ash is also a significant problem for aviation, as aircraft can lose engine power within ash clouds. On a day-to-day basis airliners are routed to take advantage of the jet stream tailwind to improve fuel efficiency. Aircrews are briefed prior to takeoff on the conditions to expect en route and at their destination. Additionally, airports often change which runway is being used to take advantage of a headwind. This reduces the distance required for takeoff, and eliminates potential crosswinds.

Marine

Commercial and recreational use of waterways can be limited significantly by wind direction and speed, wave periodicity and heights, tides, and precipitation. These factors can each influence the safety of marine transit. Consequently, a variety of codes have been established to efficiently transmit detailed marine weather forecasts to vessel pilots via radio, for example the MAFOR (marine forecast). Typical weather forecasts can be received at sea through the use of RTTY, Navtex and Radiofax.

Agriculture

Farmers rely on weather forecasts to decide what work to do on any particular day. For example, drying hay is only feasible in dry weather. Prolonged periods of dryness can ruin cotton, wheat, and corn crops. While corn crops can be ruined by drought, their dried remains can be used as a cattle feed substitute in the form of silage. Frosts and freezes play havoc with crops both during the spring and fall. For example, peach trees in full bloom can have their potential peach crop decimated by a spring freeze. Orange groves can suffer significant damage during frosts and freezes, regardless of their timing.

Forestry

Weather forecasting of wind, precipitations and humidity is essential for preventing and controlling wildfires. Different indices, like the Forest fire weather index and the Haines Index, have been developed to predict the areas more at risk to experience fire from natural or human causes. Conditions for the development of harmful insects can be predicted by forecasting the evolution of weather, too.

Utility companies

Electricity and gas companies rely on weather forecasts to anticipate demand, which can be strongly affected by the weather. They use the quantity termed the degree day to determine how strong of a use there will be for heating (heating degree day) or cooling (cooling degree day). These quantities are based on a daily average temperature of 65 °F (18 °C). Cooler temperatures force heating degree days (one per degree Fahrenheit), while warmer temperatures force cooling degree days. In winter, severe cold weather can cause a surge in demand as people turn up their heating. Similarly, in summer a surge in demand can be linked with the increased use of air conditioning systems in hot weather. By anticipating a surge in demand, utility companies can purchase additional supplies of power or natural gas before the price increases, or in some circumstances, supplies are restricted through the use of brownouts and blackouts.

Weather and Society Integrated Studies

Weather and Society Integrated Studies (WAS*IS) is an international movement that is changing the weather enterprise by integrating social science into meteorological research and practice. WAS*IS was formed to build an interdisciplinary community of practitioners, researchers and decision makers collaborating to effectively understand how to improve weather warnings, incorporate societal impacts into weather forecasts, and use social science tools and methods. WAS*IS is changing the culture from what WAS to what IS the future of integrated studies

WAS*IS has 276 representatives from the United States, Canada, China, France, Germany, Australia, New Zealand, Finland, the Netherlands, and numerous Caribbean countries. WAS*ISers come from many parts of the weather enterprise, including meteorology, emergency management, hydrology, geography, climatology, psychology, sociology, economics, ecology, education and anthropology. Workshops were held in Boulder, Colorado on August 8–15, 2008, August 6–14, 2009, August 5–13, 2010, and August 4–12, 2011. An advanced workshop was held in Norman, Oklahoma September 15–17, 2008. The first WAS*IS Caribbean was held June 6–10, 2010 in San Juan, Puerto Rico.

WAS*IS was established in 2005 at the National Center for Atmospheric Research (NCAR) located in Boulder, Colorado. WAS*IS was founded by Dr. Eve Gruntfest, a geographer at the University of Colorado at Colorado Springs, and Julie Demuth, an atmospheric scientist at NCAR, as part of NCAR's Societal Impacts Program. Dr. Gruntfest was inspired by the comments of physical scientists she met during her career who were interested in integrating meteorology and social science but were unsure how to do the work, and did not know anyone else involved in societal impacts research or practice.

WAS*IS has received support from the US Weather Research Program, the National Oceanic and Atmospheric Administration NOAA, NCAR's Societal Impacts Program, the University of Colorado at Colorado Springs, and NCAR's Institute for the Study of Society and Environment. Additional workshop support has come from the University of Oklahoma, Monash University Sustainability Institute, ARC Network for Earth System Science, Australian Bureau of Meteorology, and Emergency Management Australia.

3

Geography of Glaciers

Glaciers are massive bodies of slowly moving ice. Glaciers form on land, and they are made up of fallen snow that gets compressed into ice over many centuries. They move slowly downward from the pull of gravity. Most of the world's glaciers exist in the polar regions, in areas like Greenland, the Canadian Arctic, and Antarctica. Glaciers also can be found closer to the Equator in some mountain regions. The Andes Mountain range in South America contains some of the world's largest tropical glaciers. About 2 percent of all the water on Earth is frozen in glaciers.

Glaciers can range in age from a couple hundred to thousands of years old. Most glaciers today are remnants of the massive ice sheets that covered Earth during the Ice Age. The Ice Age ended more than 10,000 years ago. During Earth's history, there have been colder periods—when glaciers formed—and warmer periods—when glaciers melted.

Scientists who study glaciers are called glaciologists. Glaciologists began studying glaciers during the 19th century in order to look for clues about past ice ages. Today, glaciologists study glaciers for clues about global warming. Old photographs and paintings show that glaciers have melted away from mountain regions over time. Indeed, glaciers worldwide have been shrinking—and even disappearing—at an accelerated rate for the past several decades.

Among the scientists studying the changes in glaciers is Erin Christine Pettit, a glaciologist at the University of Alaska Fairbanks. Pettit observes and measures the flow, fracture, and retreat of glaciers. She uses this information to study how much water enters the oceans from melting glaciers. Melting glaciers are one factor contributing to the global sea-level rise.

A glacier is a persistent body of dense ice that is constantly moving under its own weight. A glacier forms where the accumulation of snow exceeds its ablation over many years, often centuries. Glaciers slowly deform and flow under stresses induced by their weight, creating crevasses, seracs, and other distinguishing features. They also abrade rock and debris from their substrate to create landforms such as cirques, moraines, or fjords. Glaciers form only on land and are distinct from the much thinner sea ice and lake ice that forms on the surface of bodies of water.

On Earth, 99% of glacial ice is contained within vast ice sheets (also known as "continental glaciers") in the polar regions, but glaciers may be found in mountain ranges on every continent other than the Australian mainland, including Oceania's high-latitude oceanic island countries such as New Zealand. Between latitudes 35°N and 35°S, glaciers occur only in the Himalayas, Andes, and a few high mountains in East Africa, Mexico, New Guinea and on Zard Kuh in Iran.

With more than 7,000 known glaciers, Pakistan has more glacial ice than any other country outside the polar regions.

Glaciers cover about 10% of Earth's land surface. Continental glaciers cover nearly 13 million km2 (5 million sq mi) or about 98% of Antarctica's 13.2 million km2 (5.1 million sq mi), with an average thickness of 2,100 m (7,000 ft). Greenland and Patagonia also have huge expanses of continental glaciers.

The volume of glaciers, not including the ice sheets of Antarctica and Greenland, has been estimated at 170,000 km3.

Glacial ice is the largest reservoir of fresh water on Earth, holding with ice sheets about 69 percent of the world's freshwater.

Many glaciers from temperate, alpine and seasonal polar climates store water as ice during the colder seasons and release it later in the form of meltwater as warmer summer temperatures cause the glacier to melt, creating a water source that is especially important for plants, animals and human uses when other sources may be scant. However, within high-altitude and Antarctic environments, the seasonal temperature difference is often not sufficient to release meltwater.

Since glacial mass is affected by long-term climatic changes, e.g., precipitation, mean temperature, and cloud cover, glacial mass changes are considered among the most sensitive indicators of climate change and are a major source of variations in sea level.

A large piece of compressed ice, or a glacier, appears blue, as large quantities of water appear blue. This is because water molecules absorb other colors more efficiently than blue. The other reason for the blue color of glaciers is the lack of air bubbles. Air bubbles, which give a white color to ice, are squeezed out by pressure increasing the created ice's density.

Types

Classification by size, shape and behavior

Glaciers are categorized by their morphology, thermal characteristics, and behavior. Alpine glaciers form on the crests and slopes of mountains. A glacier that fills a valley is called a valley glacier, or alternatively, an alpine glacier or mountain glacier.

A large body of glacial ice astride a mountain, mountain range, or volcano is termed an ice cap or ice field.

Ice caps have an area less than 50,000 km2 (19,000 sq mi) by definition.

Glacial bodies larger than 50,000 km2 (19,000 sq mi) are called ice sheets or continental glaciers.

Several kilometers deep, they obscure the underlying topography. Only nunataks protrude from their surfaces. The only extant ice sheets are the two that cover most of Antarctica and Greenland.

They contain vast quantities of freshwater, enough that if both melted, global sea levels would rise by over 70 m (230 ft).

Portions of an ice sheet or cap that extend into water are called ice shelves; they tend to be thin with limited slopes and reduced velocities.

Narrow, fast-moving sections of an ice sheet are called ice streams.

In Antarctica, many ice streams drain into large ice shelves. Some drain directly into the sea, often with an ice tongue, like Mertz Glacier.

Tidewater glaciers are glaciers that terminate in the sea, including most glaciers flowing from Greenland, Antarctica, Baffin, Devon, and Ellesmere Islands in Canada, Southeast Alaska, and the Northern and Southern Patagonian Ice Fields. As the ice reaches the sea, pieces break off or calve, forming icebergs. Most tidewater glaciers calve above sea level, which often results in a tremendous impact as the iceberg strikes the water. Tidewater glaciers undergo centuries-long cycles of advance and retreat that are much less affected by climate change than other glaciers.

Classification by thermal state

Thermally, a temperate glacier is at a melting point throughout the year, from its surface to its base. The ice of a polar glacier is always below the freezing threshold from the surface to its base, although the surface snowpack may experience seasonal melting. A subpolar glacier includes both temperate and polar ice, depending on the depth beneath the surface and position along the length of the glacier. In a similar way, the thermal regime of a glacier is often described by its basal temperature. A cold-based glacier is below freezing at the ice-ground interface and is thus frozen to the underlying substrate. A warm-based glacier is above or at freezing at the interface and is able to slide at this contact.

This contrast is thought to a large extent to govern the ability of a glacier to effectively erode its bed, as sliding ice promotes plucking at rock from the surface below.

Glaciers which are partly cold-based and partly warm-based are known as polythermal.

Formation

Glaciers form where the accumulation of snow and ice exceeds ablation. A glacier usually originates from a cirque landform (alternatively known as a "corrie" or as a "cwm") – a typically armchair-shaped geological feature (such as a depression between mountains enclosed by arêtes) – which collects and compresses through gravity the snow that falls into it. This snow accumulates and the weight of the snow falling above compacts it, forming névé (granular snow). Further crushing of the individual snowflakes and squeezing the air from the snow turns it into "glacial ice". This glacial ice will fill the cirque until it "overflows" through a geological weakness or vacancy, such as a gap between two mountains. When the mass of snow and ice reaches sufficient thickness, it begins to move by a combination of surface slope, gravity, and pressure. On steeper slopes, this can occur with as little as 15 m (50 ft) of snow-ice.

In temperate glaciers, snow repeatedly freezes and thaws, changing into granular ice called firn. Under the pressure of the layers of ice and snow above it, this granular ice fuses into denser firn. Over a period of years, layers of firn undergo further compaction and become glacial ice. Glacier ice is slightly more dense than ice formed from frozen water because glacier ice contains fewer trapped air bubbles.

Glacial ice has a distinctive blue tint because it absorbs some red light due to an overtone of the infrared OH stretching mode of the water molecule. (Liquid water appears blue for the same reason. The blue of glacier ice is sometimes misattributed to Rayleigh scattering of bubbles in the ice.)

In some places it is cold all year long. On the tops of mountains it can snow any time of the year. When the first snow falls there is a lot of air space between the flakes. As more and more snow falls, the snow begins to pack together and get much heavier. The heavy snow compacts and presses down on the ground.

Then when other snowstorms come the snow packs down even more, and the flakes start to lose their shape. Then the air gets sucked out of the flakes and most of the flakes turn into ice. The snowflakes get rounder as they absorb water in between the left over air spaces. As the years go by the the ice fields grow deeper and stronger until they form a glacier.

Structure

A glacier originates at a location called its glacier head and terminates at its glacier foot, snout, or terminus.

Glaciers are broken into zones based on surface snowpack and melt conditions.

The ablation zone is the region where there is a net loss in glacier mass. The upper part of a glacier, where accumulation exceeds ablation, is called the accumulation zone. The equilibrium line separates the ablation zone and the accumulation zone; it is the contour where the amount of new snow gained by accumulation is equal to the amount of ice lost through ablation. In general, the accumulation zone accounts for 60–70% of the glacier's surface area, more if the glacier calves icebergs. Ice in the accumulation zone is deep enough to exert a downward force that erodes underlying rock. After a glacier melts, it often leaves behind a bowl- or amphitheater-shaped depression that ranges in size from large basins like the Great Lakes to smaller mountain depressions known as cirques.

The accumulation zone can be subdivided based on its melt conditions.

The dry snow zone is a region where no melt occurs, even in the summer, and the snowpack remains dry.

The percolation zone is an area with some surface melt, causing meltwater to percolate into the snowpack. This zone is often marked by refrozen ice lenses, glands, and layers. The snowpack also never reaches the melting point.

Near the equilibrium line on some glaciers, a superimposed ice zone develops. This zone is where meltwater refreezes as a cold layer in the glacier, forming a continuous mass of ice.

The wet snow zone is the region where all of the snow deposited since the end of the previous summer has been raised to 0 °C.

The health of a glacier is usually assessed by determining the glacier mass balance or observing terminus behavior. Healthy glaciers have large accumulation zones, more than 60% of their area is snow-covered at the end of the melt season, and they have a terminus with a vigorous flow.

Following the Little Ice Age's end around 1850, glaciers around the Earth have retreated substantially. A slight cooling led to the advance of many alpine glaciers between 1950 and 1985, but since 1985 glacier retreat and mass loss has become larger and increasingly ubiquitous.

Motion

Glaciers move in two ways. The first way is through the pull of gravity and meltwater. Gravity pulls the heavy weight of the glacier down a hill very slowly. If you were watching one, you probably wouldn't see it moving. Under the glacier, as it slowly moves, the rocks it is dragging underneath cause the ice to melt. The water under the glacier is called melt water. The meltwater makes it slippery and helps the glacier to move down hills.

There is also meltwater on top of the glaciers that gets into cracks. When it refreezes the ice cracks and moves, kind of like the way an ice cube might crack and melt in a glass of soda. The amount of melt water in the glacier depends on the weather. Different parts of the glacier move at different speeds.

Glaciers move, or flow, downhill by the force of gravity and the internal deformation of ice.

Ice behaves like a brittle solid until its thickness exceeds about 50 m (160 ft). The pressure on ice deeper than 50 m causes plastic flow. At the molecular level, ice consists of stacked layers of molecules with relatively weak bonds between layers. When the stress on the layer above exceeds the inter-layer binding strength, it moves faster than the layer below.

Glaciers also move through basal sliding. In this process, a glacier slides over the terrain on which it sits, lubricated by the presence of liquid water. The water is created from ice that melts under high pressure from frictional heating. Basal sliding is dominant in temperate or warm-based glaciers.

Although evidence in favor of glacial flow was known by the early 19th century, other theories of glacial motion were advanced, such as the idea that meltwater, refreezing inside glaciers, caused the glacier to dilate and extend its length. As it became clear that glaciers behaved to some degree as if the ice were a viscous fluid, it was argued that "regelation", or the melting and refreezing of ice at a temperature lowered by the pressure on the ice inside the glacier, was what allowed the ice to deform and flow. James Forbes came up with the essentially correct explanation in the 1840s, although it was several decades before it was fully accepted.

Fracture zone and cracks

The top 50 m (160 ft) of a glacier are rigid because they are under low pressure. This upper section is known as the fracture zone and moves mostly as a single unit over the plastic-flowing lower section. When a glacier moves through irregular terrain, cracks called crevasses develop in the fracture zone. Crevasses form because of differences in glacier velocity. If two rigid sections of a glacier move at different speeds or directions, shear forces cause them to break apart, opening a crevasse. Crevasses are seldom more than 46 m (150 ft) deep but, in some cases, can be at least 300 m (1,000 ft) deep. Beneath this point, the plasticity of the ice prevents the formation of cracks. Intersecting crevasses can create isolated peaks in the ice, called seracs.

Crevasses can form in several different ways. Transverse crevasses are transverse to flow and form where steeper slopes cause a glacier to accelerate. Longitudinal crevasses form semi-parallel to flow where a glacier expands laterally. Marginal crevasses form near the edge of the glacier, caused by the reduction in speed caused by friction of the valley walls. Marginal crevasses are largely transverse to flow. Moving glacier ice can sometimes separate from the stagnant ice above, forming a bergschrund. Bergschrunds resemble crevasses but are singular features at a glacier's margins. Crevasses make travel over glaciers hazardous, especially when they are hidden by fragile snow bridges.

Below the equilibrium line, glacial meltwater is concentrated in stream channels. Meltwater can pool in proglacial lakes on top of a glacier or descend into the depths of a glacier via moulins. Streams within or beneath a glacier flow in englacial or sub-glacial tunnels. These tunnels sometimes reemerge at the glacier's surface.

Glacier Life

Do you ever wonder if anything lives on or in a glacier? The top of a glacier is only rocks, ice, dirt, and snow, so who or what would want to live there? Well actually, there is life on a glacier. The reason why is because of the wind. When the wind blows over the ground it collects dust and other things like insects, pollen, minerals, and bacteria that are in the air. Then when the wind blows over the glacier it drops everything on it.

Snow fleas and ice worms live on a glacier. The wind brings them their food. Ice worms are related to the earth worm, but much smaller. They are at less than an inch long!! They move around by squeezing in between the ice crystals at the top of the glacier. They can go as deep as six feet. Ice worms lay their eggs and hatch in the ice. Sometimes you can find over 100 ice worms in one area while another area won't have any.

Not many people get lucky enough to see ice worms and some people think they don't exist. Ice worms only live if the temperature is below 40 degrees Fahrenheit. If the ice gets below 22 degrees Fahrenheit, the ice worms will freeze. Ice worms only live in a glacier if it is beside the ocean or has a lot of meltwater. Ice worms eat algae, that is near the top of the glacier. Snow algae is red instead of green like most algae.

Along with the ice worm, some land insects like spiders and flies might live on a glacier. They usually get eaten by animals that are bigger, like the birds. Animals other the than ice worm also use glaciers. Tidewater glaciers give a home or resting place to large animals. Seals climb up onto the icebergs to have their babies. The new born babies, or pups, and their mothers can rest on the icebergs safe from their predators. Eagles like to stand on icebergs to look for food.

Types of Glaciers

Glaciers are categorized in many ways including by their morphology, thermal characteristics or their behavior. Alpine glaciers form on the crests and slopes of mountains and are also known as "mountain glaciers", "niche glaciers", or "cirque glaciers". An alpine glacier that fills a valley is sometimes called a valley glacier. Larger glaciers that cover an entire mountain, mountain range, or volcano are known as an ice cap or ice field, such as the Juneau Icefield. Ice caps feed outlet glaciers, tongues of ice that extend into valleys below far from the margins of the larger ice masses.

The largest glacial bodies, ice sheets or continental glaciers, cover more than 50,000 km^2 (20,000 $mile^2$). Several kilometers deep, they obscure the underlying topography. Only nunataks protrude from the surface. The only extant ice sheets are the two that cover most ofAntarctica and Greenland. These regions contain vast quantities of fresh water. The volume of ice is so large that if the Greenland ice sheetmelted, it would cause sea levels to rise six meters (20 ft) all around the world. If the Antarctic ice sheet melted, sea levels would rise up to 65 meters (210 ft). Ice shelves are areas of floating ice, commonly located at the margin of an ice sheet. As a result they are thinner and have limited slopes and reduced velocities. Ice streams are fast-moving sections of an ice sheet. They can be several hundred kilometers long. Ice streams have narrow margins and on either side ice flow is usually an order of magnitude less. In Antarctica, many ice streams drain into large ice shelves. However, some drain directly into the sea, often with an ice tongue, like Mertz Glacier. In Greenland and Antarctica ice streams ending at the sea are often referred to as tidewater glaciers or outlet glaciers, such as Jakobshavn Isbræ (Kalaallisut: Sermeq Kujalleq).

Tidewater glaciers are glaciers that terminate in the sea. As the ice reaches the sea pieces break off, or calve, forming icebergs. Most tidewater glaciers calve above sea level, which often results in a tremendous splash as the iceberg strikes the water. If the water is deep, glaciers can calve underwater, causing the iceberg to suddenly leap up out of the water.

The Hubbard Glacier is the longest tidewater glacier in Alaska and has a calving face over 10 km (6.2 mi) long. Yakutat Bay and Glacier Bay are both popular with cruise ship passengers because of the huge glaciers descending hundreds of feet to the water. This glacier type undergoes centuries-long cycles of advance and retreat that are much less affected by the climate changes currently causing the retreat of most other glaciers. Most tidewater glaciers are outlet glaciers of ice caps and ice fields.

In terms of thermal characteristics, a temperate glacier is at melting point throughout the year, from its surface to its base. The ice of a polarglacier is always below freezing point from the surface to its base, although the surface snowpack may experience seasonal melting. A sub-polar glacier has both temperate and polar ice, depending on the depth beneath the surface and position along the length of the glacier.

Glaciers in Alaska

Did you know that no one has ever counted all of the glaciers in Alaska? There are nearly 100,000 of them, but most of them don't have names. Most of the glaciers are in the southern part of Alaska. About anywhere in Alaska where you drive or go boating, you can see the remains of the ice age and movement of glaciers.

Snow Bridges

A crevasse is an open break or cut in the surface of the glacier. They are dangerous because you can't always see them. It looks like part of the surface, but it really is a deep hole. Snow bridges form on top of crevasses.

Anatomy

The location where a glacier originates is referred to as the "glacier head". A glacier terminates at the "glacier foot", or terminus. Glaciers are broken into zones based on surface snowpack and melt conditions. The ablation zone is the region where there is a net loss in glacier mass. The equilibrium line separates the ablation zone and the accumulation zone. At this altitude, the amount of new snow gained by accumulation is equal to the amount of ice lost through ablation. The accumulation zone is the region where snowpack or superimposed ice accumulation persists.

A further zonation of the accumulation zone distinguishes the melt conditions that exist.

1. The dry snow zone is a region where no melt occurs, even in the summer, and the snowpack remains dry.
2. The percolation zone is an area with some surface melt, causing meltwater to percolate into the snowpack. This zone is often marked by refrozen ice lenses, glands, and layers. The snowpack also never reaches melting point.
3. Near the equilibrium line on some glaciers, a superimposed ice zone develops. This zone is where meltwater refreezes as a cold layer in the glacier, forming a continuous mass of ice.
4. The wet snow zone is the region where all of the snow deposited since the end of the previous summer has been raised to 0 °C.

The upper part of a glacier that receives most of the snowfall is called the accumulation zone. In general, the glacier accumulation zone accounts for 60-70% of the glacier's surface area, more if the glacier calves icebergs. The depth of ice in the accumulation zone exerts a downward force sufficient to cause deep erosion of the rock in this area. After the glacier is gone, its force often leaves a bowl or amphitheater-shaped isostatic depression ranging from large lake basins, such as the Great Lakes or Finger Lakes, to smaller mountain basins, known ascirques.

The "health" of a glacier is usually assessed by determining the glacier mass balance or observing terminus behavior. Healthy glaciers have large accumulation zones, more than 60% of their area snowcovered at the end of the melt season, and a terminus with vigorous flow.

Following the Little Ice Age, around 1850, the glaciers of the Earth have retreated substantially through the 1940s. A slight cooling led to the advance of many alpine glaciers from 1950-1985. However, since 1985 glacier retreat and mass balance loss has become increasingly ubiquitous and large.

Just The Facts

- There were about 11 different ice ages.
- The ice ages were during the earth's 4.6 billion years of history.
- The last ice age was called "The Great Ice Age" and was 11,000 years ago.
- During the "Great Ice Age" over a third of the earth was covered in ice. During the ice age the air had less carbon dioxide in it.
- Right now we are living in a mini ice age.
- There are two explanations of why the ice ages might have occurred: 1.The temperatures were much colder so it never rained, only snowed. 2. The earth changed its tilt away from the sun.

Motion

Glaciers move, or flow, downhill due to the internal deformation of ice and gravity. Ice behaves like an easily breaking solid until its thickness exceeds about 50 meters (160 ft). The pressure on ice deeper than that depth causes plastic flow. At the molecular level, ice consists of stacked layers of molecules with relatively weak bonds between the layers. When the stress of the layer above exceeds the inter-layer binding strength, it moves faster than the layer below.

Another type of movement is through basal sliding. In this process, the glacier slides over the terrain on which it sits, lubricated by the presence of liquid water. As the pressure increases toward the base of the glacier, the melting point of water decreases, and the ice melts. Friction between ice and rock and geothermal heat from the Earth's interior also contribute to melting. This type of movement is dominant in temperate, or warm-based glaciers. The geothermal heat flux becomes more important the thicker a glacier becomes.

The rate of movement is dependent on the underlying slope, amongst many other factors.

Fracture zone and Cracks

The top 50 meters of the glacier, being under less pressure, are more rigid; this section is known as the fracture zone, and mostly moves as a single unit, over the plastic-like flow of the lower section. When the glacier moves through irregular terrain, cracks up to 50 meters deep form in the fracture zone. The lower layers of glacial ice flow and deform plastically under the pressure, allowing the glacier as a whole to move slowly like a viscous fluid. Glaciers flow downslope, usually this reflects the slope of their base, but it may reflect the surface slope instead. Thus, a glacier can flow rises in terrain at their base.

The upper layers of glaciers are more brittle, and often form deep cracks known as crevasses. The presence of crevasses is a sure sign of a glacier. Moving ice-snow of a glacier is often separated from a mountain side or snow-ice that is stationary and clinging to that mountain side by a bergshrund. This looks like a crevasse but is at the margin of the glacier and is a singular feature.

Crevasses form due to differences in glacier velocity. As the parts move at different speeds and directions, shear forces cause the two sections to break apart, opening the crack of a crevasse all along the disconnecting faces. Hence, the distance between the two separated parts, while touching and rubbing deep down, frequently widens significantly towards the surface layers, many times creating a wide chasm. Intersecting crevasses may create isolated peaks in the ice, called a serac.

Crevasses seldom are more than 150 feet (46 m) deep but in some cases can be 1,000 feet (300 m) or even deeper. Beneath this point, the plastic deformation of the ice under pressure is too great for the differential motion to generate cracks. Transverse crevasses are transverse to flow, as a glacier accelerates where the slope steepens. Longitudinal crevasses form semi-parallel to flow where a glacier expands laterally. Marginal crevasses form from the edge of the glacier, due to the reduction in speed caused by friction of the valley walls. Marginal crevasses are usually largely transverse to flow.

Crevasses make travel over glaciers hazardous. Subsequent heavy snow may form fragile snow bridges, increasing the danger by hiding the presence of crevasses at the surface. Below the equilibrium line, glacier meltwater is concentrated in stream channels. The meltwater can pool in a proglacial lake, a lake on top of the glacier, or can descend into the depths of the glacier via moulins. Within or beneath the glacier, the stream will flow in an englacial or sub-glacial tunnel. Sometimes these tunnels reemerge at the surface of the glacier.

Speed

The speed of glacial displacement is partly determined by friction. Friction makes the ice at the bottom of the glacier move more slowly than the upper portion. In alpine glaciers, friction is also generated at the valley's side walls, which slows the edges relative to the center. This was confirmed by experiments in the 19th century, in which stakes were planted in a line across an alpine glacier, and as time passed, those in the center moved farther.

Mean speeds vary greatly. There may be no motion in stagnant areas, where trees can establish themselves on surface sediment deposits such as in Alaska. In other cases they can move as fast as 20–30 meters per day, as in the case of Greenlands's Jakobshavn Isbræ(Kalaallisut: Sermeq Kujalleq), or 2–3 m per day on Byrd Glacier, the largest glacier in Antarctica. Velocity increases with increasing slope, increasing thickness, increasing snowfall, increasing longitudinal confinement, increasing basal temperature, increasing meltwater production and reduced bed hardness.

A few glaciers have periods of very rapid advancement called surges. These glaciers exhibit normal movement until suddenly they accelerate, then return to their previous state. During these surges, the glacier may reach velocities far greater than normal speed. These surges may be caused by failure of the underlying bedrock, the ponding of meltwater at the base of the glacier — perhaps delivered from a supraglacial lake — or the simple accumulation of mass beyond a critical "tipping point".

In glaciated areas where the glacier moves faster than one kilometer per year, glacial earthquakes occur. These are large scale tremblors that have seismic magnitudes as high as 6.1.

The number of glacial earthquakes in Greenland show a peak every year in July, August and September, and the number is increasing over time. In a study using data from January 1993 through October 2005, more events were detected every year since 2002, and twice as many events were recorded in 2005 as there were in any other year. This increase in the numbers of glacial earthquakes in Greenland may be a response to global warming.

Seismic waves are also generated by the Whillans Ice Stream, a large, fast-moving river of ice pouring from the West Antarctic Ice Sheet into the Ross Ice Shelf. Two bursts of seismic waves are released every day, each one equivalent to a magnitude 7 earthquake, and are seemingly related to the tidal action of the Ross Sea. During each event a 96 by 193 kilometer (60 by 120 mile) region of the glacier moves as much as .67 meters (2.2 ft) over about 25 minutes, remains still for 12 hours, then moves another half-meter. The seismic waves are recorded atseismographs around Antarctica, and even as far away as Australia, a distance of more than 6,400 kilometers. Because the motion takes place of such along period of time 10 to 25 minutes, it cannot be felt by scientists standing on the moving glacier. It is not known if these events are related to global warming

Ogives

These are alternating dark and light bands of ice occurring as narrow wave crests and wave valleys on glacier surfaces. They only occur below icefalls, but not all icefalls have ogives below them. Once formed, they bend progressively downglacier due to the increased velocity toward the glacier's centerline. Ogives are linked to seasonal motion of the glacier as the width of one dark and one light band generally equals the annual movement of the glacier.

The ridges and valleys are formed because ice from an icefall is severely broken up, thereby increasing ablation surface area during the summertime. This creates a swale and space for snow accumulation in the winter, which in turn creates a ridge. Sometimes ogives are described as either wave ogives or band ogives, in which they are solely undulations or varying color bands, respectively.

Geography

Glaciers are known on every continent and approximately fifty countries, a count excluding those (Australia, South Africa) that have glaciers only on distant subantarctic island territories. Extensive glaciers are found in Antarctica, Chile, Canada, Alaska, Greenland and Iceland. Mountain glaciers are widespread, e.g., in the Andes, the Himalaya, the Rocky Mountains, the Caucasus, and the Alps. On mainland Australia no glaciers exist today, although a small glacier on Mount Kosciuszko was present in the last glacial period, and Tasmania was extensively glaciated. In New Guinea, small, rapidly diminishing, glaciers are located on its highest summit massif of Puncak Jaya. Africa has glaciers on Mount Kilimanjaro in Tanzania, on Mount Kenya and in the Ruwenzori Range. The South Island of New Zealand has many glaciers including Tasman, Fox and Franz Josef Glaciers.

Among oceanic islands glaciers occur today on Iceland, Svalbard, Jan Mayen and the subantarctic islands of Marion, Heard, Grande Terre and Bouvet. During glacial periods of the Quaternary, Taiwan, Hawaii on Mauna Kea and Tenerife also had large alpine glaciers, whilst the Faroeand Crozet Islands were completely glaciated.

Permanent snow cover is affected by factors such as the degree of slope on the land, amount of snowfall and the winds. Glaciers can be found in all latitudes except from 20° to 27° north and south of the equator where the presence of the descending limb of the Hadley circulationlowers precipitation so much that with high insolation snow lines reach above 6,500 metres (21,330 ft). Between 19ÚN and 19ÚS, however, precipitation is higher and the mountains above 5,000 metres (16,400 ft) usually have permanent snow. The only snow to occur exactly on the Equator is at 4,690 m (15,387 ft) on the southern slope of Volcán Cayambe in Ecuador, whilst the nearest glacier to either Tropic is on Iztaccíhuatl in Mexico about 470 kilometres (290 mi) south of the Tropic of Cancer.

Conversely, areas of the Arctic, such as Banks Island, and the McMurdo Dry Valleys in Antarctica are considered polar deserts, as they receive little snowfall despite the bitter cold. Cold air unlike warm air is unable to transport much water vapor. Even during glacial periods of the Quaternary, Manchuria lowland Siberia, and central and northern Alaska, though extraordinarily cold had such light snowfall that glaciers could not form.

In addition to the dry, unglaciated polar regions, some mountains and volcanoes in Bolivia, Chile and Argentina are high (4,500 metres (14,800 ft) - 6,900 m (22,600 ft)) and cold, but the relative lack of precipitation prevents snow from accumulating into glaciers. This is because these peaks are located near or in the hyperarid Atacama desert.

Glacial Geology

As the glacier flows over the bedrock's fractured surface, it softens and lifts blocks of rock that are brought into the ice. This process is known as plucking, and it is produced when subglacial water penetrates the fractures and the subsequent freezing expansion separates them from the bedrock. When the ice expands, it acts as a lever that loosens the rock by lifting it. This way, sediments of all sizes become part of the glacier's load. The rocks frozen into the bottom of the ice then act like grit insandpaper.

Abrasion occurs when the ice and the load of rock fragments slide over the bedrock and function as sandpaper that smooths and polishes the surface situated below. This pulverized rock is called rock flour. The flour is formed by rock grains of a size between 0.002 and 0.00625 mm. Sometimes the amount of rock flour produced is so high that currents of meltwaters acquire a grayish color. These processes of erosion lead to steeper valley walls and mountain slopes in alpine settings, which can cause avalanches and rock slides. These further add material to the glacier.

Visible characteristics of glacial abrasion are glacial striations. These are produced when the bottom's ice contains large chunks of rock that mark scratches in the bedrock. By mapping the direction of the flutes, researchers can determine the direction of the glacier's movement. Chatter marks are seen as lines of roughly crescent-shape depressions in the rock underlying a glacier, caused by the abrasion where a boulder in the ice catches and is then released repetitively as the glacier drags it over the underlying basal rock.

The rate of glacier erosion is variable. The differential erosion undertaken by the ice is controlled by six important factors:

- Velocity of glacial movement;
- Thickness of the ice;
- Shape, abundance and hardness of rock fragments contained in the ice at the bottom of the glacier;
- Relative ease of erosion of the surface under the glacier;
- Thermal conditions at the glacier base; and
- Permeability and water pressure at the glacier base.

Material that becomes incorporated in a glacier are typically carried as far as the zone of ablation before being deposited. Glacial deposits are of two distinct types:

- Glacial till: material directly deposited from glacial ice. Till includes a mixture of undifferentiated material ranging from clay size to boulders, the usual composition of a moraine.
- Fluvial and outwash: sediments deposited by water. These deposits are stratified through various processes, such as boulders' being separated from finer particles.

The larger pieces of rock which are encrusted in till or deposited on the surface are called "glacial erratics". They may range in size from pebbles to boulders, but as they may be moved great distances, they may be of drastically different type than the material upon which they are found. Patterns of glacial erratics provide clues of past glacial motions.

Moraines

Glacial moraines are formed by the deposition of material from a glacier and are exposed after the glacier has retreated. These features usually appear as linear mounds of till, a non-sorted mixture of rock, gravel and boulders within a matrix of a fine powdery material. Terminal or end moraines are formed at the foot or terminal end of a glacier. Lateral moraines are formed on the sides of the glacier. Medial moraines are formed when two different glaciers, flowing in the same direction, coalesce and the lateral moraines of each combine to form a moraine in the middle of the merged glacier.

Less apparent is the ground moraine, also called glacial drift, which often blankets the surface underneath much of the glacier downslope from the equilibrium line. Glacial meltwaters contain rock flour, an extremely fine powder ground from the underlying rock by the glacier's movement. Other features formed by glacial deposition include long snake-like ridges formed by streambeds under glaciers, known as eskers, and distinctive streamlined hills, known as drumlins.

Stoss-and-lee erosional features are formed by glaciers and show the direction of their movement. Long linear rock scratches (that follow the glacier's direction of movement) are called glacial striations, and divots in the rock are called chatter marks. Both of these features are left on the surfaces of stationary rock that were once under a glacier and were formed when loose rocks and boulders in the ice were transported over the rock surface. Transport of fine-grained material within a glacier can smooth or polish the surface of rocks, leading to glacial polish. Glacial erratics are rounded boulders that were left by a melting glacier and are often seen perched precariously on exposed rock faces after glacial retreat.

The term moraine is of French origin. It was coined by peasants to describe alluvial embankments and rims found near the margins of glaciers in the French Alps. In modern geology, the term is used more broadly, and is applied to a series of formations, all of which are composed of till.

Drumlins

Drumlins are asymmetrical, canoe shaped hills with aerodynamic profiles made mainly of till. Their heights vary from 15 to 50 meters and they can reach a kilometer in length. The tilted side of the hill looks toward the direction from which the ice advanced (stoss), while the longer slope follows the ice's direction of movement (lee).

Drumlins are found in groups called drumlin fields or drumlin camps. An example of these fields is found east of Rochester, New York, and it is estimated that it contains about 10,000 drumlins.

Although the process that forms drumlins is not fully understood, it can be inferred from their shape that they are products of the plastic deformation zone of ancient glaciers. It is believed that many drumlins were formed when glaciers advanced over and altered the deposits of earlier glaciers.

Glacial Valleys

Before glaciation, mountain valleys have a characteristic "V" shape, produced by downward erosion by water. However, during glaciation, these valleys widen and deepen, forming a "U"-shaped glacial valley. Besides the deepening and widening of the valley, the glacier also smooths the valley due to erosion. In this way, it eliminates the spurs of earth that extend across the valley. Because of this interaction, triangular cliffs called truncated spurs are formed.

Many glaciers deepen their valleys more than their smaller tributaries. Therefore, when the glaciers recede from the region, the valleys of the tributary glaciers remain above the main glacier's depression, and these are called hanging valleys.

In parts of the soil that were affected by abrasion and plucking, the depressions left can be filled by lakes, called paternoster lakes.

At the 'start' of a classic valley glacier is the cirque, which has a bowl shape with escarped walls on three sides, but open on the side that descends into the valley. In the cirque, an accumulation of ice is formed. These begin as irregularities on the side of the mountain, which are later augmented in size by the coining of the ice. Once the glacier melts, these corries are usually occupied by small mountain lakes calledtarns.

There may be two glacial cirques 'back to back' which erode deep into their backwalls until only a narrow ridge, called an arête is left. This structure may result in a mountain pass.

Glaciers are also responsible for the creation of fjords (deep coves or inlets) and escarpments that are found at high latitudes.

Arêtes and Horns (pyramid peak)

An arête is a narrow crest with a sharp edge. The meeting of three or more arêtes creates pointed pyramidal peaks and in extremely steep-sided forms these are called horns.

Both features may have the same process behind their formation: the enlargement of cirques from glacial plucking and the action of the ice. Horns are formed by cirques that encircle a single mountain.

Arêtes emerge in a similar manner; the only difference is that the cirques are not located in a circle, but rather on opposite sides along a divide. Arêtes can also be produced by the collision of two parallel glaciers. In this case, the glacial tongues cut the divides down to size through erosion, and polish the adjacent valleys.

Roche Moutonnée

Some rock formations in the path of a glacier are sculpted into small hills with a shape known as roche moutonnée or "sheepback" rock. An elongated, rounded, asymmetrical, bedrock knob can be produced by glacier erosion. It has a gentle slope on its up-glacier side and a steep to vertical face on the down-glacier side. The glacier abrades the smooth slope that it flows along, while rock is torn loose from the downstream side and carried away in ice, a process known as 'plucking'. Rock on this side is fractured by a combination of various forces, such as water, ice in rock cracks, and structural stresses.

Alluvial Stratification

The water that rises from the ablation zone moves away from the glacier and carries with it fine eroded sediments. As the speed of the water decreases, so does its capacity to carry objects in suspension. The water then gradually deposits the sediment as it runs, creating an alluvial plain. When this phenomenon occurs in a valley, it is called a valley train. When the deposition is to an estuary, the sediments are known as "bay mud".

Outwash plains and valley trains are usually accompanied by basins known as "kettles". These are glacial depressions produced when large ice blocks are stuck in the glacial alluvium. After they melt, the sediment is left with holes. The diameter of such depressions ranges from 5 m to 13 km, with depths of up to 45 meters. Most are circular in shape due to the melting blocks of ice becoming rounded. The lakes that often form in these depressions are known as "kettle lakes".

Glacial Deposits

When a glacier reduces in size to a critical point, its flow stops, and the ice becomes stationary. Meanwhile, meltwater flows over, within, and beneath the ice leave stratified alluvial deposits. Because of this, as the ice melts, it leaves stratified deposits in the form of columns, terraces and clusters. These types of deposits are known as "glacial deposits".

When those deposits take the form of hills or mounds, they are called kames. Some kames form when meltwater deposits sediments through openings in the interior of the ice. In other cases, they are just the result of fans ordeltas towards the exterior of the ice produced by meltwater. When the glacial ice occupies a valley, it can form terraces or kame along the sides of the valley.

A third type of deposit formed in contact with the ice is characterized by long, narrow sinuous crests, composed fundamentally of sand and gravel deposited by streams of meltwater flowing within, or beneath the glacier. After the ice has melted, these linear ridges or eskers remain as landscape features. Some of these crests have heights exceeding 100 meters and their lengths surpass 100 km.

Loess Deposits

Very fine glacial sediments or rock flour is often picked up by wind blowing over the bare surface and may be deposited great distances from the original fluvial deposition site. These eolian loess deposits may be very deep, even hundreds of meters, as in areas of China and the Midwestern United States of America. Katabatic winds can be important in this process.

Transportation and Erosion

- Entrainment is the picking up of loose material by the glacier from along the bed and valley sides. Entrainment can happen by regelation or by the ice simply picking up the debris.
- Basal ice freezing is thought to be to be made by glaciohydraulic supercooling, though some studies show that even where physical conditions allow it to occur, the process may not be responsible for observed sequences of basal ice.
- Plucking is the process involves the glacier freezing onto the valley sides and subsequent ice movement pulling away masses of rock. As the bedrock is greater

in strength than the glacier, only previously loosened material can be removed. It can be loosened by local pressure and temperature, water and pressure release of the rock itself.

- Supraglacial debris is carried on the surface of the glacier as lateral and medial moraines. In summer ablation, surface melt water carries a small load and this often disappears down crevasses.
- Englacial debris is moraine carried within the body of the glacier.
- Subglacial debris is moved along the floor of the valley either by the ice as ground moraine or by meltwater streams formed by pressure melting.

Deposition

- Lodgement till is identical to ground moraine. It is material that is smeared on to the valley floor when its weight becomes too great to be moved by the glacier.
- Ablation till is a combination of englacial and supraglacial moraine. It is released as a stationary glacier begins to melt and material is dropped in situ.
- Dumping is when a glacier moves material to its outermost or lowermost end and dumps it.
- Deformation flow is the change of shape of the rock and land due to the glacier.

Isostatic Rebound

This rise of a part of the crust is due to an isostatic adjustment. A large mass, such as an ice sheet/glacier, depresses the crust of the Earth and displaces the mantle below. The depression is about a third the thickness of the ice sheet. After the glacier melts the mantle begins to flow back to its original position pushing the crust back to its original position. This post-glacial rebound, which lags melting of the ice sheet/ glacier, is currently occurring in measurable amounts in Scandinavia and the Great Lakes region of North America.

An interesting geomorphological feature created by the same process, but on a smaller scale, is known as dilation-faulting. It occurs within rock where previously compressed rock is allowed to return to its original shape, but more rapidly than can be maintained without faulting, leading to an effect similar to that which would be seen if the rock were hit by a large hammer. This can be observed in recently de-glaciated parts of Iceland and Cumbria.

Glacial Erosion

PictureGlaciers erode the underlying rock by abrasion and plucking. Glacial meltwater seeps into cracks of the underlying rock, the water freezes and pushes pieces of rock outward. The rock is then plucked out and carried away by the flowing ice of the moving glacier. With the weight of the ice over them, these rocks can scratch deeply into the underlying bedrock making long, parallel grooves in the bedrock, called glacial striations. Mountain glaciers leave behind unique erosional features. When a glacier cuts through a 'V' shaped river valley, the glacier pucks rocks from the sides and bottom. This widens the valley and steepens the walls, making a 'U' shaped valley.For example, Utah's landscape has been greatly influenced by glacial activity of the last ice age. Using Google Maps for our map, first look at Little Cottonwood Canyon and notice how linear the canyon is. That is a direct result of an alpine glacier carving down creating a U-shaped canyon that ultimately ended in Lake Bonneville. Now using the same map, compare Little Cottonwood Canyon with Big Cottonwood Canyon. If you were to drive from the Salt Lake Valley up the canyon, you would notice the canyon to be more V-shaped since because it was carved out by the stream and spring runoff instead of a glacier. The canyon is also more jagged in shape rather than linear like U-shaped canyons. But notice about half way up the canyon, the direction of the canyon changes from the northeast to running southeast. At that pivot point is where the canyon stopped being carved by a river and was carved by another alpine glacier. From that pivot point to the end of the canyon, the canyon because linear and more U-shaped. The cirque, where the glacier originated, is at the head of the canyon is now home to Solitude Sky Resort and a portion of the melted cirque is now called Silver Lake.Smaller tributary glaciers, like tributary streams, flow into the main glacier in their own shallower 'U' shaped valleys. A hanging valley forms where the main glacier cuts off a tributary glacier and creates a cliff. Streams plunge over the cliff to create waterfalls. Up high on a mountain, where a glacier originates, rocks are pulled away from valley walls.

Depositional Features of Glaciers

As glaciers flow, mechanical weathering loosens rock on the valley walls, which falls as debris on the glacier. Glaciers can carry rocks of any size, from giant boulders to silt. These rocks can be carried for many miles over many years and decades. These rocks that are different in type or origin from the surrounding bedrock are glacial erratics. Melting glaciers deposit all the big and small bits of rocky material they are carrying in a pile. These unsorted deposits of rock are called glacial till. Glacial till is found in different types of deposits. Linear rock deposits are called moraines and are named by their location relative to the glacier. Geologists study moraines to figure out how far glaciers extended and how long it took them to melt away.

Medial Moraine

PictureLateral moraines form at the edges of the glacier as material drops onto the glacier from erosion of the valley walls. Medial moraines form where the lateral moraines of two tributary glaciers join together in the middle of a larger glacier. Sediment from underneath the glacier becomes a ground moraine after the glacier melts. Ground moraine contributes to the fertile transported soils in many regions.

Terminal moraines are long ridges of till left at the furthest point the glacier reached. End moraines are deposited where the glacier stopped for a long enough period to create a rocky ridge as it retreated. Long Island in New York is formed by two end moraines.

While glaciers dump unsorted sediments, glacial melt water can sort and re-transport the sediments. As water moves through unsorted glacial till, it leaves behind the larger particles and takes away the smaller bits of sand and silt.Several types of stratified deposits form in glacial regions but are not formed directly by the ice. Varves form where lakes are covered by ice in the winter. Dark, fine-grained clays sink to the bottom in winter but melting ice in spring brings running water that deposits lighter colored sands. Each alternating dark/light layer represents one year of deposits. If during a year, a glacier accumulates more ice than melts away, the glacier advances downhill. If a glacier melts more than it accumulates over a year, it is retreating.

Climate Change and Glacial Loss

There has been a lot of conversations and research in the scientific community about the dramatic amount of glacial melt that has been occurring around the planet as of late. Glaciers flow incredibly slowly down slope under the influence of gravity. But satellite imagery and on the ground data have shown that these glaciers are melting back and receding at an incredible rate. Because glaciers take so long to form and flow, they are excellent indicators of the temperature of the planet. As the planet has continued to warm over the last 100 years, glaciers from all around the world are retreating back. This is important to humanity because glaciers make up a large portion of the usable fresh water on the planet. As these glaciers melt away, so do huge amounts of fresh drinking water, just at a time when the human population continues to grow at an exponential rate.

4

Soil, Rivers and Streams: Physical Geography

Soil, also commonly referred to as earth or dirt, is a mixture of organic matter, minerals, gases, liquids, and organisms that together support life. (Some scientific definitions distinguish dirt from soil by restricting the former term specifically to displaced soil.) The term pedolith, used commonly to refer to the soil, translates to ground stone in the sense fundamental stone, from the ancient Greek word p?d??, meaning 'ground, earth'.

Soil consists of a solid phase of minerals and organic matter (the soil matrix), as well as a porous phase that holds gases (the soil atmosphere) and water (the soil solution).

Accordingly, soil is a three-state system of solids, liquids, and gases.

Soil is a product of several factors: the influence of climate, relief (elevation, orientation, and slope of terrain), organisms, and the soil's parent materials (original minerals) interacting over time.

It continually undergoes development by way of numerous physical, chemical and biological processes, which include weathering with associated erosion. Given its complexity and strong internal connectedness, soil ecologists regard soil as an ecosystem.

Most soils have a dry bulk density (density of soil taking into account voids when dry) between 1.1 and 1.6 g/cm3, though the soil particle density is much higher, in the range of 2.6 to 2.7 g/cm3.

Little of the soil of planet Earth is older than the Pleistocene and none is older than the Cenozoic, although fossilized soils are preserved from as far back as the Archean.

The pedosphere interfaces with the lithosphere, the hydrosphere, the atmosphere, and the biosphere.

Collectively, Earth's body of soil, called the pedosphere, has four important functions:

as a medium for plant growth

as a means of water storage, supply and purification as a modifier of Earth's atmosphere as a habitat for organisms

All of these functions, in their turn, modify the soil and its properties.

Soil science has two basic branches of study: edaphology and pedology. Edaphology studies the influence of soils on living things.

Pedology focuses on the formation, description (morphology), and classification of soils in their natural environment.

In engineering terms, soil is included in the broader concept of regolith, which also includes other loose material that lies above the bedrock, as can be found on the Moon and other celestial objects.

Processes

Soil is a major component of the Earth's ecosystem. The world's ecosystems are impacted in far-reaching ways by the processes carried out in the soil, with effects ranging from ozone depletion and global warming to rainforest destruction and water pollution. With respect to Earth's carbon cycle, soil acts as an important carbon reservoir, and it is potentially one of the most reactive to human disturbance and climate change.

As the planet warms, it has been predicted that soils will add carbon dioxide to the atmosphere due to increased biological activity at higher temperatures, a positive feedback (amplification).

This prediction has, however, been questioned on consideration of more recent knowledge on soil carbon turnover.

Soil acts as an engineering medium, a habitat for soil organisms, a recycling system for nutrients and organic wastes, a regulator of water quality, a modifier of atmospheric composition, and a medium for plant growth, making it a critically important provider of ecosystem services.

Since soil has a tremendous range of available niches and habitats, it contains a prominent part of the Earth's genetic diversity. A gram of soil can contain billions of organisms, belonging to thousands of species, mostly microbial and largely still unexplored.

Soil has a mean prokaryotic density of roughly 108 organisms per gram, whereas the ocean has no more than 107 prokaryotic organisms per milliliter (gram) of seawater.

Organic carbon held in soil is eventually returned to the atmosphere through the process of respiration carried out by heterotrophic organisms, but a substantial part is retained in the soil in the form of soil organic matter; tillage usually increases the rate of soil respiration, leading to the depletion of soil organic matter.

Since plant roots need oxygen, aeration is an important characteristic of soil. This ventilation can be accomplished via networks of interconnected soil pores, which also absorb and hold rainwater making it readily available for uptake by plants. Since plants require a nearly continuous supply of water, but most regions receive sporadic rainfall, the water-holding capacity of soils is vital for plant survival. Soils can effectively remove impurities, kill disease agents and degrade contaminants, this latter property being called natural attenuation.

Typically, soils maintain a net absorption of oxygen and methane and undergo a net release of carbon dioxide and nitrous oxide.

Soils offer plants physical support, air, water, temperature moderation, nutrients, and protection from toxins.

Soils provide readily available nutrients to plants and animals by converting dead organic matter into various nutrient forms.

Composition

A typical soil is about 50% solids (45% mineral and 5% organic matter), and 50% voids (or pores) of which half is occupied by water and half by gas.

The percent soil mineral and organic content can be treated as a constant (in the short term), while the percent soil water and gas content is considered highly variable whereby a rise in one is simultaneously balanced by a reduction in the other.

The pore space allows for the infiltration and movement of air and water, both of which are critical for life existing in soil.

Compaction, a common problem with soils, reduces this space, preventing air and water from reaching plant roots and soil organisms.

Given sufficient time, an undifferentiated soil will evolve a soil profile which consists of two or more layers, referred to as soil horizons. These differ in one or more properties such as in their texture, structure, density, porosity, consistency, temperature, color, and reactivity.

The horizons differ greatly in thickness and generally lack sharp boundaries; their development is dependent on the type of parent material, the processes that modify those parent materials, and the soil-forming factors that influence those processes. The biological influences on soil properties are strongest near the surface, though the geochemical influences on soil properties increase with depth. Mature soil profiles typically include three basic master horizons: A, B, and C. The solum normally includes the A and B horizons. The living component of the soil is largely confined to the solum, and is generally more prominent in the A horizon.

It has been suggested that the pedon, a column of soil extending vertically from the surface to the underlying parent material and large enough to show the characteristics of all its horizons, could be subdivided in the humipedon (the living part, where most soil organisms are dwelling, corresponding to the humus form), the copedon (in intermediary position, where most weathering of minerals takes place) and the lithopedon (in contact with the subsoil).

The soil texture is determined by the relative proportions of the individual particles of sand, silt, and clay that make up the soil. The interaction of the individual mineral particles with organic matter, water, gases via biotic and abiotic processes causes those particles to flocculate (stick together) to form aggregates or peds.

Where these aggregates can be identified, a soil can be said to be developed, and can be described further in terms of color, porosity, consistency, reaction (acidity), etc.

Water is a critical agent in soil development due to its involvement in the dissolution, precipitation, erosion, transport, and deposition of the materials of which a soil is composed.

The mixture of water and dissolved or suspended materials that occupy the soil pore space is called the soil solution. Since soil water is never pure water, but contains hundreds of dissolved organic and mineral substances, it may be more accurately called the soil solution. Water is central to the dissolution, precipitation and leaching of minerals from the soil profile. Finally, water affects the type of vegetation that grows in a soil, which in turn affects the development of the soil, a complex feedback which is exemplified in the dynamics of banded vegetation patterns in semi-arid regions.

Soils supply plants with nutrients, most of which are held in place by particles of clay and organic matter (colloids)

The nutrients may be adsorbed on clay mineral surfaces, bound within clay minerals (absorbed), or bound within organic compounds as part of the living organisms or dead soil organic matter. These bound nutrients interact with soil water to buffer the soil solution composition (attenuate changes in the soil solution) as soils wet up or dry out, as plants take up nutrients, as salts are leached, or as acids or alkalis are added.

Plant nutrient availability is affected by soil pH, which is a measure of the hydrogen ion activity in the soil solution. Soil pH is a function of many soil forming factors, and is generally lower (more acid) where weathering is more advanced.

Most plant nutrients, with the exception of nitrogen, originate from the minerals that make up the soil parent material. Some nitrogen originates from rain as dilute nitric acid and ammonia, but most of the nitrogen is available in soils as a result of nitrogen fixation by bacteria. Once in the soil-plant system, most nutrients are recycled through living organisms, plant and microbial residues (soil organic matter), mineral-bound forms, and the soil solution. Both living soil organisms (microbes, animals and plant roots) and soil organic matter are of critical importance to this recycling, and thereby to soil formation and soil fertility. Microbial soil enzymes may release nutrients from minerals or organic matter for use by plants and other microorganisms, sequester (incorporate) them into living cells, or cause their loss from the soil by volatilisation (loss to the atmosphere as gases) or leaching.

SOIL-FORMING MATERIALS

Rocks are the chief sources for the parent materials over which soils are developed. There are three main kinds of rocks:

(i) igneous rocks,
(ii) sedimentary rocks, and
(iii) metamorphic rocks.

Igneous rocks. They are formed by the cooling, hardening and crystallizing of various kinds of lavas and differ widely in their chemical composition. They chiefly contain feldspars, maphic minerals and quartz. Rocks containing a high proportion of quartz (60-75%) are classified as acidic, whereas those containing less than 50%

quartz are classified as basic. The common igneous rocks found in India are the granites(acidic) and basalts or the Deccan Trap (basic)

Sedimentary rocks. They are derived from igneous rocks and are formed by the consolidation of fragmentary rock materials and the products of their decomposition deposited by water. The common sedimentary rocks are conglomerate, sandstone, shale and limestone. Alluvial, glacial and aeolian deposits form the unconsolidated sedimentary rocks.

Metamorphic rocks. They are formed from the igneous or sedimentary rocks by the action of intense heat and high pressure or both resulting in considerable change in the texture and mineral composition. The common metamorphic rocks are gneis from granite, quartzite from quartz or sandstone, marble from limestone and slate from shale.

History of the Study of Soil

The history of the study of soil is intimately tied to our urgent need to provide food for ourselves and forage for our animals. Throughout history, civilizations have prospered or declined as a function of availability and productivity of soils.

The Greek historian Xenophon (450-355 B.C.) is credited with being the first to expound upon the merits of green-manuring crops, "But then whatever weeds are upon the ground, being turned into earth, enrich the soil as much a dung."

Columella's Husbandry, circa 60 A.D. was used by 15 generations (450 years) of those encompassed by the Roman Empire until its collapse. From the fall of Rome to the French Revolution, knowledge of soil and agriculture was passed on from parent to child and as a result, crop yields were low. During the Dark Ages for Europe, Yahya Ibn_al-'Awwam's handbook guided the people of North Africa, Spain and the Middle East with its emphasis on irrigation, a translation of which was finally carried to the southwest of the United States.

Jethro Tull, an English gentleman, introduced in 1701 an improved grain drill that systemized the planting of seed and invented a horse-drawn weed hoe, the two of which allowed fields once choked with weeds to be brought back to production and seed to be used more economically. Tull, however, also introduced the mistaken idea that manure introduced weed seeds, and that fields should be plowed in order to pulverize the soil and so release the locked up nutrients. His ideas were taken up and carried to their extremes in the 20th century, when farmers repeatedly plowed fields far beyond what was necessary to control weeds. During a period of drought, the repeated plowing resulted in the Dust Bowl in the prairie region of the Central United States and Canada.

The "two-course system" of a year of wheat followed by a year of fallow was replaced in the 18th century by the Norfolk four-course system, in which wheat was grown in the first year, turnips the second, followed by barley, with clover and ryegrass together, in the third. The taller barley was harvested in the third year while the clover and ryegrass were grazed or cut for feed in the fourth. The turnips fed cattle and sheep in the winter. The fodder crops produced large supplies of animal manure, which returned nutrients to the soil.

Experiments into what made plants grow first led to the idea that the ash left behind when plant matter was burnt was the essential element and overlooked the role of nitrogen, which is not left on the ground after combustion. Jan Baptist van Helmont thought he had proved water to be the essential element from his famous experiment with a willow tree grown in carefully controlled conditions in which only water was added, which after five years of growth was removed and weighed, roots and all, and found to weigh 165 pounds. The oven-dried soil, originally 200 pounds, was again dried and weighed and found to have lost only two ounces, which van Helmont reasonably explained as experimental error and assumed that the soil had in fact lost nothing. As rain water was the only thing added by the experimenter, he concluded that water was the essential element in plant life. In fact the two ounces lost from the soil were the minerals taken up by the willow tree during its growth.

John Woodward experimented with various types of water ranging from clean to muddy and found muddy water the best, and so he concluded earthy matter was the essential element. Others concluded it was humus in the soil that passed some essence to the growing plant.

The French chemist Antonine Lavoisier showed that plants and animals must "combust" oxygen internally to live and was able to deduce that most of the 165-pound weight of Van Helmont's willow tree derived from air. The chemical basis of nutrients delivered to the soil in manure was emphasized and in the mid-19th century chemical fertilizers were used, but the dynamic interaction of soil and its life forms awaited discovery.

It was known that nitrogen was essential for growth and in 1880 the presence of Rhizobium bacteria in the roots of legumes explained the increase of nitrogen in soils so cultivated. The importance of life forms in soil was finally recognized.

Crop rotation, mechanization, chemical and natural fertilizers led to a doubling of wheat yields in western Europe between 1800 to 1900.

Soils of India

Types of Soil Found in India

Indian soils are generally divided into four broad types. These soil types are: 1) alluvial soils; 2) regur soils; 3) red soils and 4) laterite soils.

ALLUVIAL SOILS: This is the most important and widespread category. It covers 40% of the land area. In fact the entire Northern Plains are made up of these soils. They have been brought down and deposited by three great Himalayan rivers- Sutlej, Ganga and Brahmaputra- and their tributaries. Through a narrow corridor in Rajasthan they extend into the plains of Gujarat. They are common in eastern coastal plains and in the deltas of Mahanadi, Godavari, Krishna and Kaveri.

REGUR SOILS: These soils are black in colour and are also known as black soils. Since, they are ideal for growing cotton, they are also called cotton soils, in addition to their normal nomenclature of regur soils. These soils are most typical of the Deccan trap (Basalt) region spread over north-west Deccan plateau and are made up of lava flows. They cover the plateaus of Mahrashtra, Saurashtra, Malwa and southern Madhya Pradesh and extends eastwards in the south along the Godavari and Krishna Valleys.

RED SOILS: These soils are developed on old crystalline rocks under moderate to heavy rainfall conditions. They are deficient in phosphoric acid, organic matter and nitrogenous material. Red soils cover the eastern part of the peninsular region comprising Chhotanagpur plateau, Orissa, eastern Madhya Pradesh, Telangana, the Nilgiris and Tamil Nadu plateau. Tey extended northwards in the west along the Konkan coast of Maharashtra.

LATERITE SOILS: The laterite soils is the result of intense leaching owing to heavy tropical rains. They are found along the edge of plateau in the east covering small parts of Tamil Nadu, and Orissa and a small part of Chhotanagpur in the north and Meghalaya in the north-east.

Soil Forming Factors

Soil formation, or pedogenesis, is the combined effect of physical, chemical, biological, and anthropogenic processes on soil parent material. Soil genesis involves processes that develop layers or horizons in the soil profile. These processes involve additions, losses, transformations and translocations of material that compose the soil. Minerals derived fromweathered rocks undergo changes that cause the formation of secondary minerals and other compounds that are variably soluble in water. These constituents are moved (translocated) from one area of the soil to other areas by water and animal activity. The alteration and movement of materials within soil causes the formation of distinctive soil horizons.

How the soil "life" cycle proceeds is influenced by at least five classic soil forming factors that are dynamically intertwined in shaping the way soil is developed: parent material, regionalclimate, topography, biotic potential and the passage of time. An example of soil development would begin with the weathering of lava flow bedrock which would produce the purely mineral-based parent material from which soils form. Soil development would start from bare rock of recent flows in warm regions under heavy and very frequent rainfall. In such climates, plants become established very quickly on basaltic lava, even though there is very little organic material.

The plants are supported by the porous rock as it is filled with nutrient-bearing water that carries dissolved minerals from rocks and guano. Crevasses and pockets, local topography of the rocks, would hold fine materials and harbor plant roots. The developing plant roots themselves are associated with mycorrhizal fungi that gradually break up the porous lava, and by these means organic matter and a finer mineral soil accumulate with time.

Parent Material

The material from which soil forms is called parent material. It includes: weathered primary bedrock; secondary material transported from other locations, namely colluvium and alluvium; deposits that are already present but mixed or altered in other ways—old soil formations, organic material; and anthropogenic materials, such as landfill or mine waste.

Soils that develop from their underlying parent rocks are called "residual soils", and have the same general chemistry as their parent rocks. Few soils form in such a manner.

Most soils derive from transported parent materials that have been moved many miles by wind, water and gravity. . Windblown material called loess, common in the Midwest of North America and in Central Asia, may have been moved many hundreds of miles.

Cumulose parent material includes peats and mucks and may develop in place from plant residues that have been preserved by the low oxygen content of a high water table.

Weathering is the first stage in the transforming of parent material into soil material. In soils forming from bedrock, a thick layer of weathered material called saprolite may form.

Physical properties

The physical properties of soils, in order of decreasing importance for ecosystem services such as crop production, are texture, structure, bulk density, porosity, consistency, temperature, colour and resistivity. Soil texture is determined by the relative proportion of the three kinds of soil mineral particles, called soil separates: sand, silt, and clay. At the next larger scale, soil structures called peds or more commonly soil aggregates are created from the soil separates when iron oxides, carbonates, clay, silica and humus, coat particles and cause them to adhere into larger, relatively stable secondary structures. Soil bulk density, when determined at standardized moisture conditions, is an estimate of soil compaction. Soil porosity consists of the void part of the soil volume and is occupied by gases or water. Soil consistency is the ability of soil materials to stick together. Soil temperature and colour are self-defining. Resistivity refers to the resistance to conduction of electric currents and affects the rate of corrosion of metal and concrete structures which are buried in soil. These properties vary through the depth of a soil profile, i.e. through soil horizons. Most of these properties determine the aeration of the soil and the ability of water to infiltrate and to be held within the soil.

Soil moisture

Soil water content can be measured as volume or weight. Soil moisture levels, in order of decreasing water content, are saturation, field capacity, wilting point, air dry, and oven dry. Field capacity describes a drained wet soil at the point water content reaches equilibrium with gravity. Irrigating soil above field capacity risks percolation losses. Wilting point describes the dry limit for growing plants.

Available water capacity is the amount of water held in a soil profile available to plants. As water content drops, plants have to work against increasing forces of adherence and sorptivity to withdraw water. Irrigation scheduling avoids moisture stress by replenishing depleted water before stress is induced.

Capillary action is responsible for moving groundwater from wet regions of the soil to dry areas. Subirrigation designs (e.g., wicking beds, sub-irrigated planters) rely on capillarity to supply water to plant roots. Capillary action can result in an evaporative concentration of salts, causing land degradation through salination.

Soil moisture measurement–measuring the water content of the soil, as can be expressed in terms of volume or weight–can be based on in situ probes (e.g., capacitance probes, neutron probes), or remote sensing methods.

Soil gas

The atmosphere of soil, or soil gas, is very different from the atmosphere above. The consumption of oxygen by microbes and plant roots, and their release of carbon dioxide, decreases oxygen and increases carbon dioxide concentration. Atmospheric CO2 concentration is 0.04%, but in the soil pore space it may range from 10 to 100 times that level, thus potentially contributing to the inhibition of root respiration. Calcareous soils regulate CO2 concentration by carbonate buffering, contrary to acid soils in which all CO2 respired accumulates in the soil pore system. At extreme levels, CO2 is toxic. This suggests a possible negative feedback control of soil CO2 concentration through its inhibitory effects on root and microbial respiration (also called 'soil respiration'). In addition, the soil voids are saturated with water vapour, at least until the point of maximal hygroscopicity, beyond which a vapour-pressure deficit occurs in the soil pore space.

Adequate porosity is necessary, not just to allow the penetration of water, but also to allow gases to diffuse in and out. Movement of gases is by diffusion from high concentrations to lower, the diffusion coefficient decreasing with soil compaction. Oxygen from above atmosphere diffuses in the soil where it is consumed and levels of carbon dioxide in excess of above atmosphere diffuse out with other gases (including greenhouse gases) as well as water. Soil texture and structure strongly affect soil porosity and gas diffusion. It is the total pore space (porosity) of soil, not the pore size, and the degree of pore interconnection (or conversely pore sealing), together with water content, air turbulence and temperature, that determine the rate of diffusion of gases into and out of soil. Platy soil structure and soil compaction (low porosity) impede gas flow, and a deficiency of oxygen may encourage anaerobic bacteria to reduce (strip oxygen) from nitrate NO3 to the gases N2, N2O, and NO, which are then lost to the atmosphere, thereby depleting the soil of nitrogen, a detrimental process called denitrification. Aerated soil is also a net sink of methane (CH4) but a net producer of methane (a strong heat-absorbing greenhouse gas) when soils are depleted of oxygen and subject to elevated temperatures.

Soil atmosphere is also the seat of emissions of volatiles other than carbon and nitrogen oxides from various soil organisms, e.g. roots, bacteria, fungi, animals. These volatiles are used as chemical cues, making soil atmosphere the seat of interaction networks playing a decisive role in the stability, dynamics and evolution of soil ecosystems. Biogenic soil volatile organic compounds are exchanged with the aboveground atmosphere, in which they are just 1–2 orders of magnitude lower than those from aboveground vegetation.

Humans can get some idea of the soil atmosphere through the well-known 'after-the-rain' scent, when infiltering rainwater flushes out the whole soil atmosphere after a drought period, or when soil is excavated, a bulk property attributed in a reductionist manner to particular biochemical compounds such as petrichor or geosmin.

Solid phase (soil matrix)

Soil particles can be classified by their chemical composition (mineralogy) as well as their size. The particle size distribution of a soil, its texture, determines many of the properties of that soil, in particular hydraulic conductivity and water potential, but the mineralogy of those particles can strongly modify those properties. The mineralogy of the finest soil particles, clay, is especially important.

Chemistry

For the academic discipline, see Soil chemistry.

The chemistry of a soil determines its ability to supply available plant nutrients and affects its physical properties and the health of its living population. In addition, a soil's chemistry also determines its corrosivity, stability, and ability to absorb pollutants and to filter water. It is the surface chemistry of mineral and organic colloids that determines soil's chemical properties. A colloid is a small, insoluble particle ranging in size from 1 nanometer to 1 micrometer, thus small enough to remain suspended by Brownian motion in a fluid medium without settling. Most soils contain organic colloidal particles called humus as well as the inorganic colloidal particles of clays. The very high specific surface area of colloids and their net electrical charges give soil its ability to hold and release ions. Negatively charged sites on colloids attract and release cations in what is referred to as cation exchange. Cation-exchange capacity is the amount of exchangeable cations per unit weight of dry soil and is expressed in terms of milliequivalents of positively charged ions per 100 grams of soil (or centimoles of positive charge per kilogram of soil; cmolc/kg). Similarly, positively charged sites on colloids can attract and release anions in the soil, giving the soil anion exchange capacity.

Cation and anion exchange

The cation exchange, that takes place between colloids and soil water, buffers (moderates) soil pH, alters soil structure, and purifies percolating water by adsorbing cations of all types, both useful and harmful.

The negative or positive charges on colloid particles make them able to hold cations or anions, respectively, to their surfaces. The charges result from four sources.

Isomorphous substitution occurs in clay during its formation, when lower-valence cations substitute for higher-valence cations in the crystal structure. Substitutions in the outermost layers are more effective than for the innermost layers, as the electric charge strength drops off as the square of the distance. The net result is oxygen atoms with net negative charge and the ability to attract cations.

Edge-of-clay oxygen atoms are not in balance ionically as the tetrahedral and octahedral structures are incomplete.

Hydroxyls may substitute for oxygens of the silica layers, a process called hydroxylation. When the hydrogens of the clay hydroxyls are ionised into solution, they leave the oxygen with a negative charge (anionic clays).

Hydrogens of humus hydroxyl groups may also be ionised into solution, leaving, similarly to clay, an oxygen with a negative charge.

Cations held to the negatively charged colloids resist being washed downward by water and are out of reach of plant roots, thereby preserving the fertility of soils in areas of moderate rainfall and low temperatures.

There is a hierarchy in the process of cation exchange on colloids, as cations differ in the strength of adsorption by the colloid and hence their ability to replace one another (ion exchange). If present in equal amounts in the soil water solution:

Al3+ replaces H+ replaces Ca2+ replaces Mg2+ replaces K+ same as NH+

4 replaces Na+

If one cation is added in large amounts, it may replace the others by the sheer force of its numbers. This is called law of mass action. This is largely what occurs with the addition of cationic fertilisers (potash, lime).

As the soil solution becomes more acidic (low pH, meaning an abundance of H+), the other cations more weakly bound to colloids are pushed into solution as hydrogen ions occupy exchange sites (protonation). A low pH may cause the hydrogen of hydroxyl groups to be pulled into solution, leaving charged sites on the colloid available to be occupied by other cations. This ionisation of hydroxy groups on the surface of soil colloids creates what is described as pH-dependent surface charges. Unlike permanent charges developed by isomorphous substitution, pH-dependent charges are variable and increase with increasing pH. Freed cations can be made available to plants but are also prone to be leached from the soil, possibly making the soil less fertile. Plants are able to excrete H+ into the soil through the synthesis of organic acids and by that means, change the pH of the soil near the root and push cations off the colloids, thus making those available to the plant.

Cation exchange capacity (CEC)

Cation exchange capacity is the soil's ability to remove cations from the soil water solution and sequester those to be exchanged later as the plant roots release hydrogen ions to the solution. CEC is the amount of exchangeable hydrogen cation (H+) that will combine with 100 grams dry weight of soil and whose measure is one milliequivalents per 100 grams of soil (1 meq/100 g). Hydrogen ions have a single charge and one-thousandth of a gram of hydrogen ions per 100 grams dry soil gives a measure of one milliequivalent of hydrogen ion. Calcium, with an atomic weight 40 times that of hydrogen and with a valence of two, converts to $(40 \div 2) \times 1$ milliequivalent = 20 milliequivalents of hydrogen ion per 100 grams of dry soil or 20 meq/100 g. The modern measure of CEC is expressed as centimoles of positive charge per kilogram (cmol/kg) of oven-dry soil.

River

A river is a natural flowing watercourse, usually freshwater, flowing towards an ocean, sea, lake or another river. In some cases, a river flows into the ground and becomes dry at the end of its course without reaching another body of water. Small rivers can be referred to using names such as stream, creek, brook, rivulet, and rill. There are no official definitions for the generic term river as applied to geographic features, although in some countries or communities a stream is defined by its size. Many names for small rivers are specific to geographic location; examples are "run" in some parts of the United States, "burn" in Scotland and northeast England, and "beck" in northern England. Sometimes a river is defined as being larger than a creek, but not always: the language is vague.

Rivers are part of the hydrological cycle. Water generally collects in a river from precipitation through a drainage basin from surface runoff and other sources such as groundwater recharge, springs, and the release of stored water in natural ice and snowpacks (e.g., from glaciers).

Rivers and streams are often considered major features within a landscape; however, they actually only cover around 0.1% of the land on Earth. They are made more obvious and significant to humans since many human cities and civilizations are built around the freshwater supplied by rivers and streams.

Most of the major cities of the world are situated on the banks of rivers, as they are, or were, used as a source of water, for obtaining food, for transport, as borders, as a defensive measure, as a source of hydropower to drive machinery, for bathing, and as a means of disposing of waste. In the preindustrial era larger rivers were a major obstruction to the movement of people, goods, and armies across them. Towns often developed at the few locations they could be crossed. Many major cities such as London are located at the lowest point at which a river could be bridged.

Potamology is the scientific study of rivers, while limnology is the study of inland waters in general.

Source and drainage basin

A river begins at a source (or more often several sources) which is usually a watershed, drains all the streams in its drainage basin, follows a path called a rivercourse (or just course) and ends at either at a mouth or mouths which could be a confluence, river delta, etc. The water in a river is usually confined to a channel, made up of a stream bed between banks. In larger rivers there is often also a wider floodplain shaped by flood-waters over-topping the channel. Floodplains may be very wide in relation to the size of the river channel. This distinction between river channel and floodplain can be blurred, especially in urban areas where the floodplain of a river channel can become greatly developed by housing and industry.

The term upriver (or upstream) refers to the direction towards the source of the river, i.e. against the direction of flow. Likewise, the term downriver (or downstream) describes the direction towards the mouth of the river, in which the current flows. The term left bank refers to the left bank in the direction of flow, right bank to the right.

River channel

Rivers can flow down mountains, through valleys (depressions) or along plains, and can create canyons or gorges. The river channel typically contains a single stream of water, but some rivers flow as several interconnecting streams of water, producing a braided river.

Extensive braided rivers are now found in only a few regions worldwide, such as the South Island of New Zealand. They also occur on peneplains and some of the larger river deltas. Anastamosing rivers are similar to braided rivers and are quite rare. They have multiple sinuous channel - 1 carrying large volumes of sediment. There are rare cases of river bifurcation in which a river divides and the resultant flows ending in different seas. An example is the bifurcation of Nerodime River in Kosovo.

A river flowing in its channel is a source of energy that acts on the river channel to change its shape and form. In 1757, the German hydrologist Albert Brahms empirically observed that the submerged weight of objects that may be carried away by a river is proportional to the sixth power of the river flow speed.

This formulation is also sometimes called Airy's law.

Thus, if the speed of flow is doubled, the flow would dislodge objects with 64 times as much submerged weight. In mountainous torrential zones, this can be seen as erosion channels through hard rocks and the creation of sands and gravels from the destruction of larger rocks. A river valley that was created from a U-shaped glaciated valley, can often easily be identified by the V-shaped channel that it has carved. In the middle reaches where a river flows over flatter land, meanders may form through erosion of the river banks and deposition on the inside of bends. Sometimes the river will cut off a loop, shortening the channel and forming an oxbow lake or billabong. Rivers that carry large amounts of sediment may develop conspicuous deltas at their mouths. Rivers whose mouths are in saline tidal waters may form estuaries.

Throughout the course of the river, the total volume of water transported downstream will often be a combination of the free water flow together with a substantial volume flowing through sub-surface rocks and gravels that underlie the river and its floodplain (called the hyporheic zone). For many rivers in large valleys, this unseen component of flow may greatly exceed the visible flow.

The water in a river is usually confined to a channel, made up of a stream bed between banks. In larger rivers there is also a wider floodplain shaped by flood-waters over-topping the channel. Flood plains may be very wide in relation to the size of the river channel. This distinction between river channel and floodplain can be blurred especially in urban areas where the floodplain of a river channel can become greatly developed by housing and industry.

The term upriver refers to the direction leading to the source of the river, which is against the direction of flow. Likewise, the term downriver describes the direction towards the mouth of the river, in which the current flows

The river channel typically contains a single stream of water, but some rivers flow as several interconnecting streams of water, producing abraided river. Extensive braided rivers are now found in only a few regions worldwide, such as the South Island of New Zealand. They also occur on peneplains and some of the larger river deltas. Anastamosing rivers are similar to braided rivers and are also quite rare. They have multiple sinuous channels carrying large volumes of sediment.

A river flowing in its channel is a source of energy which acts on the river channel to change its shape and form. In 1757, the German hydrologist Albert Brahms empirically observed that the submerged weight of objects that may be carried away by a river is proportional to the sixth power of the river flow speed. (This formulation is also sometimes called Airy's law.) Thus, if the speed of flow were doubled, the flow would dislodge objects with 64 times as much submerged weight. In mountainous torrential zones this can be seen as erosion channels through hard rocks and the creation of sands and gravels from the destruction of larger rocks. In U-shaped glaciated valleys, the subsequent river valley can often easily be identified by the V-shaped channel that it has carved.

In the middle reaches where the river may flow over flatter land, meanders may form through erosion of the river banks and deposition on the inside of bends. Sometimes the river will cut off a loop, shortening the channel and forming an oxbow lake or billabong. Rivers that carry large amounts of sediment may develop conspicuous deltasat their mouths, if conditions permit. Rivers whose mouths are in saline tidal waters may form estuaries.

Throughout the course of the river, the total volume of water transported downstream will often be a combination of the free water flow together with a substantial contribution flowing through sub-surface rocks and gravels that underlie the river and its floodplain (called the hyporheic zone). For many rivers in large valleys, this unseen component of flow may greatly exceed the visible flow.

Subsurface Streams

Most but not all rivers flow on the surface. Subterranean rivers flow underground in caves or caverns. Such rivers are frequently found in regions with limestone geologic formations. Subglacial streams are the braided rivers that flow at the beds of glaciers and ice sheets, permitting meltwater to be discharged at the front of the glacier. Because of the gradient in pressure due to the overlying weight of the glacier, such streams can even flow uphill.

Permanence of Flow

An intermittent river (or ephemeral river) only flows occasionally and can be dry for several years at a time. These rivers are found in regions with limited or highly variable rainfall, or can occur because of geologic conditions such as having a highly permeable river bed. Some ephemeral rivers flow during the summer months but not in the winter. Such rivers are typically fed from chalk aquifers which recharge from winter rainfall. In the UK these rivers are called Bournes and give their name to place such as Bournemouth and Eastbourne.

Classification

Rivers have been classified by many criteria including their topography, their biotic status, their relevance to white water rafting or canoeing activities.

Topographical Classification

Rivers can generally be classified as either alluvial, bedrock, or some mix of the two. Alluvial rivers have channels and floodplains that are self-formed in unconsolidated or weakly consolidated sediments. They erode their banks and deposit material on bars and their floodplains. Bedrock rivers form when the river downcuts through the modern sediments and into the underlying bedrock.

This occurs in regions that have experienced some kind of uplift (thereby steepening river gradients) or in which a particular hard lithology causes a river to have a steepened reach that has not been covered in modern alluvium. Bedrock rivers very often contain alluvium on their beds; this material is important in eroding and sculpting the channel. Rivers that go through patches of bedrock and patches of deep alluvial cover are classified as mixed bedrock-alluvial.

Alluvial rivers can be further classified by their channel pattern as meandering, braided, wandering, anastomose, or straight. The morphology of an alluvial river reach is controlled by a combination of sediment supply, substrate composition, discharge, vegetation, and bed aggradation.

At the turn of the 20th century William Morris Davis devised the "cycle of erosion" method of classifying rivers based on their "age". Although Davis's system is still found in many books today, after the 1950s and 1960s it became increasingly criticized and rejected by geomorphologists. His scheme did not produce testable hypotheses and was therefore deemed non-scientific. Examples of Davis's river "ages" include:

- Youthful river: A river with a steep gradient that has very few tributaries and flows quickly. Its channels erode deeper rather than wider. Examples include the Brazos, Trinity and Ebro rivers.
- Mature river: A river with a gradient that is less steep than those of youthful rivers and flows more slowly. A mature river is fed by many tributaries and has more discharge than a youthful river. Its channels erode wider rather than deeper. Examples include the Mississippi, Saint Lawrence, Danube, Ohio, Thames and Paraná rivers.
- Old river: A river with a low gradient and low erosive energy. Old rivers are characterized by flood plains. Examples include the Yellow, Ganges, Tigris, Euphrates, Indus and Nilerivers.
- Rejuvenated river: A river with a gradient that is raised by tectonic uplift.

The way in which a river's characteristics vary between the upper course and lower course of a river are summarized by the Bradshaw model. Power-law relationships between channel slope, depth, and width are given as a function of discharge by "river regime".

Biotic Classification

There are very many systems of classification based on biotic conditions typically assigning classes from the most oligotrophic or unpolluted through to the most eutrophic or polluted. Other systems are based on a whole eco-system approach such as developed by the New Zealand Ministry for the Environment. In Europe, the requirements of the Water Framework Directive has led to the development of a wide range of classification methods including classifications based on fishery status A system of river zonation used in francophonecommunities divides rivers into three primary zones:

- The crenon is the uppermost zone at the source of the river. It is further divided into the eucrenon (spring or boil zone) and the hypocrenon (brook or headstream zone). These areas are characterized by low temperatures, reduced oxygen content and slow moving water.
- The rhithron is the upstream portion of the river that follows the crenon. It is characterized by relatively cool temperatures, high oxygen levels, and fast, turbulent flow.
- The potamon is the remaining downstream stretch of river. It is characterized by warmer temperatures, lower oxygen levels, slow flow and sandier bottoms.

Whitewater Classification

The International Scale of River Difficulty is used to rate the challenges of navigation—particularly those with rapids. Class I is the easiest and Class VI is the hardest.

Himalayan Rivers

The main Himalayan river systems are the Ganga, the Indus and the Brahmaputra river systems. The Himalayan rivers form large basins. Many rivers pass through the Himalayas. These deep valleys with steep rock sides were formed by the down - cutting of the river during the period of the Himalayan uplift. They perform intense erosional activity up the streams and carry huge load of sand and silt. In the plains, they form large meanders, and a variety of depositional features like flood plains, river cliffs and levees.

These rivers are perennial as they get water from the rainfall as well as the melting of ice. Nearly all of them create huge plains and are navigable over long distances of their course. These rivers are also harnessed in their upstream catchment area to generate hydroelectricity.

Peninsular Rivers

The main peninsular river systems include the Narmada, the Tapi, the Godavari, the Krishna, the Kaveri and the Mahanadi river systems. The Peninsular rivers flow through shallow valleys. A large number of them are seasonal as their flow is dependent on rainfall. The intensity of erosional activities is also comparatively low because of the gentler slope. The hard rock bed and lack of silt and sand does not allow any significant meandering. Many rivers therefore have straight and linear courses. These rivers provide huge opportunities for hydro-electric power.

The Indus River System

The Indus originates in the northern slopes of the Kailash range in Tibet near Lake Manasarovar. It follows a north-westerly course through Tibet. It enters Indian territory in Jammu and Kashmir.

It forms a picturesque gorge in this part. Several tributaries - the Zaskar, the Shyok, the Nubra and the Hunza join it in the Kashmir region. It flows through the regions of Ladakh, Baltistan and Gilgit and runs between the Ladakh Range and the Zaskar Range. It crosses the Himalayas through a 5181 m deep gorge near Attock, lying north of the Nanga Parbat and later takes a bend to the south west direction before entering Pakistan. It has a large number of tributaries in both India and Pakistan and has a total length of about 2897 km from the source to the point near Karachi where it falls into the Arabian Sea. The main tributaries of the Indus in India are Jhelum, Chenab, Ravi, Beas and Sutlej.

Jhelum

The Jhelum originates in the south-eastern part of Kashmir, in a spring at Verinag. It flows into the Wular Lake, which lies to the north, and then into Baramula. Between Baramula and Muzaffarabad it enters a deep gorge cut by the river in the Pir Panjal range. It has a right bank tributary the Kishanganga which joins it at Muzaffarabad. It follows the Indo-Pakistan border flowing into the plains of Punjab, finally joining the Chenab at Trimmu.

Chenab

The Chenab originates from the confluence of two rivers, the Chandra and the Bhaga, which themselves originate from either side of the Bara Lacha Pass in Lahul. It is also known as the Chandrabhaga in Himachal Pradesh. It runs parallel to the Pir Panjal Range in the north-westerly direction, and cuts through the range near Kishtwar. It enters the plains of Punjab near Akhnur and is later joined by the Jhelum. It is further joined by the Ravi and the Sutlej in Pakistan.

Ravi

The Ravi originates near the Rotang pass in the Kangra Himalayas and follows a north-westerly course. It turns to the south-west, near Dalhousie, and then cuts a gorge in the Dhaola Dhar range entering the Punjab plain near Madhopur. It flows as a part of the Indo-Pakistan border for some distance before entering Pakistan and joining the Chenab river. The total length of the river is about 720 km.

Beas

The Beas originates in Beas Kund, lying near the Rohtang pass. It runs past Manali and Kulu, where its beautiful valley is known as the Kulu valley. It first follows a north-west path from the town of Mandi and later a westerly path, before entering the Punjab plains near Mirthal. It joins the Sutlej river near Harika, after being joined by a few tributaries. The total length of the river is 615 km.

Sutlej

The Sutlej originates from the Rakas Lake, which is connected to the Manasarovar lake by a stream, in Tibet. Its flows in a north-westerly direction and enters Himachal Pradesh at the Shipki Pass, where it is joined by the Spiti river. It cuts deep gorges in the ranges of the Himalayas, and finally enters the Punjab plain after cutting a gorge in a hill range, the Naina Devi Dhar, where the Bhakra Dam having a large reservoir of water, called the Gobind Sagar, has been constructed. It turns west below Rupar and is later joined by the Beas. It enters Pakistan near Sulemanki, and is later joined by the Chenab. It has a total length of almost 1500 km.

The Brahmaputra River System

The Brahmaputra originates in the Mansarovar lake, also the source of the Indus and the Satluj. It is slightly longer than the Indus, but most of its course lies outside India. It flows eastward, parallel to the Himalayas. Reaching Namcha Barwa (7757 m), it takes a U-turn around it and enters India in Arunachal Pradesh and known as dihang. The undercutting done by this river is of the order of 5500 metres. In India, it flows through Arunachal Pradesh and Assam, and is joined by several tributaries.

In Tibet, the river is known as the Tsangpo. There, it receives less volume of water and has less silt. But in India, it passes through a region of heavy rainfall and as such, the river carries a large amount of rainfall and considerable amount of silt. The Brahmaputra has a braided channel throughout most of its length in Assam, with a few large islands within the channel.

The shifting of the channels of the river is also very common. The fury of the river during rains is very high. It is known for creating havoc in Assam and Bangladesh. At the same time, quite a few big pockets suffer from drought.

The Narmada River System

The Narmada or Nerbudda is a river in central India. It forms the traditional boundary between North India and South India, and is a total of 1,289 km (801 mi) long. Of the major rivers of peninsular India, only the Narmada, the Tapti and the Mahi run from east to west. It rises on the summit of Amarkantak Hill in Madhya Pradesh state, and for the first 320 kilometres (200 miles) of its course winds among the Mandla Hills, which form the head of the Satpura Range; then at Jabalpur, passing through the 'Marble Rocks', it enters the Narmada Valley between the Vindhya and Satpura ranges, and pursues a direct westerly course to the Gulf of Cambay. Its total length through the states of Madhya Pradesh, Maharashtra, and Gujarat amounts to 1312 kilometres (815 miles), and it empties into the Arabian Sea in the Bharuch district of Gujarat.

The Tapi River System

The Tapi is a river of central India. It is one of the major rivers of peninsular India with the length of around 724 km, and only the Tapi River along with the Narmada river, and the Mahi River run from east to west. It rises in the eastern Satpura Range of southern Madhya Pradesh state, and flows westward, draining Madhya Pradesh's historic Nimar region, Maharashtra's historic Khandesh and east Vidarbha regions in the northwest corner of the Deccan Plateau and South Gujarat before emptying into the Gulf of Cambay of the Arabian Sea, in the State of Gujarat. The Western Ghats or Sahyadri range starts south of the Tapti River near the border of Gujarat and Maharashtra.

The Tapi River Basin lies mostly in northern and eastern districts Maharashtra state viz, Amravati, Akola, Buldhana, Washim, Jalgaon, Dhule, Nandurbar, Malegaon, Nashik districts but also covers Betul, Burhanpur districts of Madhya Pradesh and Surat district in Gujarat as well.

The principal tributaries of Tapi River are Purna River, Girna River, Panzara River, Waghur River, Bori River and Aner River.

The Godavari River System

The river with second longest course within India, Godavari is often referred to as the Vriddh (Old) Ganga or the Dakshin (South) Ganga. The name may be apt in more ways than one, as the river follows the course of Ganga's tragedy. The river is about 1,450 km (900 miles) long.

It rises at Trimbakeshwar, near Nasik and Mumbai (formerly Bombay) in Maharashtra around 380 km distance from the Arabian Sea, but flows southeast across south-central India through the states of Madhya Pradesh, Karnataka, Orissa and Andhra Pradesh, and empties into the Bay of Bengal.

At Rajahmundry, 80 km from the coast, the river splits into two streams thus forming a very fertile delta. Like any other major rivers in India, the banks of this river also has many pilgrimage sites, Nasik, Triyambak and Bhadrachalam, being the major ones.

It is a seasonal river, widened during the monsoons and dried during the summers. Godavari river water is brownish. Some of its tributaries include Indravati River, Pranahita (Combination of Penuganga and Warda), Manjira, Bindusara and Sabari. Some important urban centers on its banks include Nasik, Bhadrachalam, Rajahmundry and Narsapur. The Asia's largest rail-cum-road bridge on the river Godavari linking Kovvur and Rajahmundry is considered to be an engineering feat.

The Krishna River System

The Krishna is one of the longest rivers of India (about 1300 km in length). It originates at Mahabaleswar in Maharashtra, passes through Sangli and meets the sea in the Bay of Bengal at Hamasaladeevi in Andhra Pradesh. The Krishna River flows through the states of Maharashtra, Karnataka and Andhra Pradesh.

The traditional source of the river is a spout from the mouth of a statue of a cow in the ancient temple of Mahadev in Mahabaleshwar.

Its most important tributary is the Tungabhadra River, which itself is formed by the Tunga and Bhadra rivers that originate in the Western Ghats. Other tributaries include the Koyna, Bhima, Mallaprabha, Ghataprabha, Yerla, Warna, Dindi, Musi and Dudhganga rivers.

The Kaveri River System

The Kaveri (also spelled Cauvery or Kavery) is one of the great rivers of India and is considered sacred by the Hindus. This river is also called Dakshin Ganga. The headwaters are in the Western Ghats range of Karnataka state, and from Karnataka through Tamil Nadu. It empties into the Bay of Bengal. Its waters have supported irrigated agriculture for centuries, and the Kaveri has been the lifeblood of the ancient kingdoms and modern cities of South India.

The source of the river is Talakaveri located in the Western Ghats about 5,000 feet (1,500 m) above sea level. Talakaveri is a famous pligrimage and tourist spot set amidst Bramahagiri Hills near Madikeri in Kodagu district of Karnataka. Thousands of piligrims flock to the temple at the source of the river especially on the specified day known as Tula sankramana when the river water has been witnessed to gush out like a fountain at a predetermined time.

It flows generally south and east for around 765 km, emptying into the Bay of Bengal through two principal mouths. Its basin is estimated to be 27,700 square miles (71,700 km^2), and it has many tributaries including Shimsha, Hemavati, Arkavathy, Kapila, Honnuhole, Lakshmana Tirtha, Kabini, Lokapavani, Bhavani, Noyyal and Famous Amaravati.

The Mahanadi River System

The Mahanadi is a river of eastern India. The Mahanadi rises in the Satpura Range of central India, and flows east to the Bay of Bengal. The Mahanadi drains most of the state of Chhattisgarh and much of Orissa and also Jharkhand and Maharashtra. It has a length of about 860 km.

Stream Order Classification

The Strahler Stream Order ranks rivers based on the connectivity and hierarchy of contributing tributaries. Headwaters are first order while the Amazon River is twelfth order. Approximately 80% of the rivers and streams in the world are of the first and second order.

Uses

Rivers have been used as a source of water, for obtaining food, for transport, as a defensive measure, as a source of hydropower to drive machinery, for bathing, and as a means of disposing of waste.

Rivers have been used for navigation for thousands of years. The earliest evidence of navigation is found in the Indus Valley Civilization, which existed in northwestern

Pakistan around 3300 BC. Riverine navigation provides a cheap means of transport, and is still used extensively on most major rivers of the world like the Amazon, the Ganges, the Nile, the Mississippi, and the Indus. Since river boats are often not regulated, they contribute a large amount to global greenhouse gas emissions, and to local cancer due to inhaling of particulates emitted by the transports.

In some heavily forested regions such as Scandinavia and Canada, lumberjacks use the river to float felled trees downstream to lumber camps for further processing, saving much effort and cost by transporting the huge heavy logs by natural means.

Rivers have been a source of food since pre-history. They can provide a rich source of fish and other edible aquatic life, and are a major source of fresh water, which can be used for drinking and irrigation. It is therefore no surprise to find most of the major cities of the world situated on the banks of rivers. Rivers help to determine the urban form of cities and neighbourhoods and their corridors often present opportunities for urban renewal through the development of foreshore ways such as river walks. Rivers also provide an easy means of disposing of waste-water and, in much of the less developed world, other wastes.

Fast flowing rivers and waterfalls are widely used as sources of energy, via watermills and hydroelectric plants. Evidence of watermills shows them in use for many hundreds of years such as in Orkney at Dounby click mill. Prior to the invention of steam power, water-mills for grinding cereals and for processing wool and other textiles were common across Europe. In the 1890s the first machines to generate power from river water were established at places such as Cragside in Northumberland and in recent decades there has been a significant increase in the development of large scale power generation from water, especially in wet mountainous regions such as Norway.

The coarse sediments, gravel, and sand, generated and moved by rivers are extensively used in construction. In parts of the world this can generate extensive new lake habitats as gravel pits re-fill with water. In other circumstances it can destabilise the river bed and the course of the river and cause severe damage to spawning fish populations which rely on stable gravel formations for egg laying.

In upland rivers, rapids with whitewater or even waterfalls occur. Rapids are often used for recreation, such as whitewater kayaking. Rivers have been important in determining political boundaries and defending countries. For example, the Danube was a long-standing border of the Roman Empire, and today it forms most of the border between Bulgaria and Romania. The Mississippi in North America and the Rhine in Europe are major east-west boundaries in those continents. The Orange and Limpopo Rivers in southern Africa form the boundaries between provinces and countries along their routes.

Ecosystem

The organisms in the riparian zone respond to changes in river channel location and patterns of flow. The ecosystem of rivers is generally described by the River continuum concept, which has some additions and refinements to allow for spatial (dams, waterfalls) and temporal (extensive flooding). The basic idea is that the river can be described as a system that is continuously changing along its length in the physical parameters, the availability of food particles and the composition of the ecosystem. The food (energy) that is the leftover of the upstream part is being utilized downstream.

The general pattern is that the first order streams contain particulate matter (decaying leaves from the surrounding forests), which is processed there by shredders like Plecoptera larvae. The leftovers of the shredders are utilized by collectors, such as Hydropsyche, and further downstream algae that create the primary production become the main foodsource of the organisms. All changes are gradual and the distribution of each species can be described as a normal curve with the highest density where the conditions are optimal. In riverssuccession is virtually absent and the composition of the ecosystem stays fixed in time.

Chemistry

The chemistry of rivers is complex and depends on inputs from the atmosphere, the geology through which it travels and the inputs from man's activities. The chemistry of the water has a large impact on the ecology of that water for both plants and animals and it also affects the uses that may be made of the river water. Understanding and characterising river water chemistry requires a well designed and managed sampling and analysis.

Like many other Aquatic ecosystems, rivers too are under increasing threat of pollution. According to a study of the WWF's Global Freshwater Programme the 10 most polluted rivers are: Ganges Indus, Yangtze, Salween-Nu, Mekong-Lancang, Rio Grande/Rio Bravo Río de la Plata, Danube, Nile-Lake Victori and the Murray-Darling.

Flow

Studying the flows of rivers is one aspect of hydrology.

Direction

A common misconception is that most, or even all, rivers flow from north to south. Rivers in fact flow downhill regardless of compass direction. Sometimes downhill is from north to south, but equally it can be from south to north, and usually is a complex meandering path involving all directions of the compass. Three of the ten longest rivers in the world—the Nile, Yenisei, and Ob—flow north, as do other major rivers such as the Rhine, Mackenzie, and Nelson.

Rivers flowing downhill, from river source to river mouth, do not necessarily take the shortest path. For alluvial streams, straight and braided rivers have very low sinuosity and flow directly down hill, while meandering rivers flow from side to side across a valley. Bedrock rivers typically flow in either afractal pattern, or a pattern that is determined by weaknesses in the bedrock, such as faults, fractures, or more erodible layers.

Rate

Volumetric flow rate, also called discharge, volume flow rate, and rate of water flow, is the volume of water which passes through a given cross-section of the river channel per unit time. It is typically measured in cubic meters per second (cumec) or cubic feet per second (cfs), where 1 m3/s = 35.51 ft3/s; it is sometimes also measured in litres or gallons per second.

Volumetric flow rate can be thought of as the mean velocity of the flow through a given cross-section, times that cross-sectional area. Mean velocity can be approximated through the use of the Law of the Wall. In general, velocity increases with the depth (or hydraulic radius) and slope of the river channel, while the cross-sectional area scales with the depth and the width: the double-counting of depth shows the importance of this variable in determining the discharge through the channel.

Sediment Yield

Sediment yield is the total quantity of particulate matter (suspended or bedload) reaching the outlet of a drainage basin over a fixed time frame. Yield is usually expressed as kilograms per square kilometre per year. Sediment delivery processes are affected by a myriad of factors such as drainage area size, basin slope, climate, sediment type (lithology), vegetation cover, and human land use / management practices.

The theoretical concept of the 'sediment delivery ratio' (ratio between yield and total amount of sediment eroded) captures the fact that not all of the sediment eroded within a catchment reaches the outlet (due to, for example, deposition on floodplains). Such storage opportunities are typically increased in catchments of larger size, thus leading to a lower yield and sediment delivery ratio.

Management

Rivers are often managed or controlled to make them more useful, or less disruptive, to human activity.

- Dams or weirs may be built to control the flow, store water, or extract energy.
- Levees, known as dikes in Europe, may be built to prevent river water from flowing on floodplains or floodways.
- Canals connect rivers to one another for water transfer or navigation.
- River courses may be modified to improve navigation, or straightened to increase the flow rate.

River management is a continuous activity as rivers tend to 'undo' the modifications made by people. Dredged channels silt up, sluice mechanisms deteriorate with age, levees and dams may suffer seepage or catastrophic failure. The benefits sought through managing rivers may often be offset by the social and economic costs of mitigating the bad effects of such management. As an example, in parts of the developed world, rivers have been confined within channels to free up flat flood-plain land for development. Floods can inundate such development at high financial cost and often with loss of life.

Rivers are increasingly managed for habitat conservation, as they are critical for many aquatic and riparian plants, resident and migratory fishes, waterfowl, birds of prey, migrating birds, and many mammals.

Stream

A stream is a body of water with a current, confined within a bed and stream banks. Depending on its locale or certain characteristics, a stream may be referred to as a branch, brook, beck, burn, creek, "crick", gill (occasionally ghyll), kill, lick, rill, river, syke, bayou, rivulet, streamage, wash, run or runnel.

Streams are important as conduits in the water cycle, instruments in groundwater recharge, and corridors for fish and wildlife migration. The biological habitat in the immediate vicinity of a stream is called a riparian zone. Given the status of the ongoing Holocene extinction, streams play an important corridor role in connecting fragmented habitats and thus in conserving biodiversity. The study of streams andwaterways in general is known as surface hydrology and is a core element of environmental geography.

River

A large natural stream, which may be a waterway.

Creek

- In North America, Australia and New Zealand, a small to medium sized natural stream. Sometimes navigable by motor craft and may be intermittent.
- In parts of New England, the UK and India, a tidal inlet, typically in a salt marsh or mangrove swamp, or between enclosed and drained former salt marshes or swamps (e.g. Port Creek separating Portsea Island from the mainland). In these cases, the stream is the tidal stream, the course of the seawater through the creek channel at low and high tide.

Tributary

A contributory stream, or a stream which does not reach the sea but joins another river (a parent river). Sometimes also called a branch or fork.

Brook

A stream smaller than a creek, especially one that is fed by a spring or seep. It is usually small and easily forded. A brook is characterized by its shallowness and its bed being composed primarily of rocks.

Runnel

The linear channel between the parallel ridges or bars on a shoreline beach or river floodplain, or between a bar and the shore. Also called aswale.

Other Names

In the United Kingdom, there are several regional names for a stream:

- Beck is used in Yorkshire, Lancashire, Dumfriesshire and Cumbria.
- Bourne is used in the chalk downland of southern England (although strictly a bourne is wet in summer and dry in winter).
- Brook is used in the Midlands, Lancashire and Cheshire.
- Burn is used in Scotland and North East England.
- Nant is used in Wales.
- Stream is used in Southern England.
- Syke is used in lowland Scotland and Cumbria.
- Allt is used in Highland Scotland.
- In North America:
- Kill in southern New York, Pennsylvania, Delaware, and New Jersey comes from a Dutch language word meaning "riverbed" or "water channel", and can also be used for the UK meaning of 'creek'.
- Run in Ohio, Pennsylvania, Maryland, or Virginia can be the name of a stream.
- Branch, fork, or prong can refer to tributaries or distributaries that share the same name as the main stream, generally with the addition of a cardinal direction.
- Branch is also used to name streams in Maryland and Virginia.
- Falls is also used to name streams in Maryland. Little Gunpowder Falls and The Jones Falls are actually rivers named in this manner, unique to Maryland.
- Stream and brook are used in Midwestern states, Mid-Atlantic states and New England.
- Crick is used in some parts of the United States.

Sources

Streams typically derive most of their water from precipitation in the form of rain and snow. Most of this water re-enters the atmosphere by evaporationfrom soil and water bodies, or by the evapotranspiration of plants. Some of the water proceeds to sink into the earth by infiltration and becomesgroundwater, much of which eventually enters streams.

Some precipitated water is temporarily locked up in snow fields and glaciers, to be released later by evaporation or melting. The rest of the water flows off the land as runoff, the proportion of which varies according to many factors, such as wind, humidity, vegetation, rock types, and relief. This runoff starts as a thin film called sheet wash, combined with a network of tiny rills, together constituting sheet runoff; when this water is concentrated in a channel, a stream has its birth.

Characteristics

Ranking

To qualify as a stream it must be either recurring or perennial. Recurring (intermittent) streams have water in the channel for at least part of the year. A stream of the first order is a stream which does not have any other recurring or perennial stream feeding into it. When two first-order streams come together, they form a second-order stream. When two second-order streams come together, they form a third-order stream.

Streams of lower order joining a higher order stream do not change the order of the higher stream. Thus, if a first-order stream joins a second-order stream, it remains a second-order stream. It is not until a second-order stream combines with another second-order stream that it becomes a third-order stream.

The gradient of a stream is a critical factor in determining its character and is entirely determined by its base level of erosion. The base level of erosion is the point at which the stream either enters the ocean, a lake or pond, or enters a stretch in which it has a much lower gradient, and may be specifically applied to any particular stretch of a stream.

In geologic terms, the stream will erode down through its bed to achieve the base level of erosion throughout its course. If this base level is low, then the stream will rapidly cut through underlying strata and have a steep gradient, and if the base level is relatively high, then the stream will form a flood plain and meander.

Meander

Meanders are looping changes of direction of a stream caused by the erosion and deposition of bank materials. These are typically serpentine in form. Typically, over time the meanders gradually migrate downstream.

If some resistant material slows or stops the downstream movement of a meander, a stream may erode through the neck between two legs of a meander to become temporarily straighter, leaving behind an arc-shaped body of water termed an oxbow lake or bayou. A flood may also cause a meander to be cut through in this way.

Typically, streams are said to have a particular profile, beginning with steep gradients, no flood plain, and little shifting of channels, eventually evolving into streams with low gradients, wide flood plains, and extensive meanders. The initial stage is sometimes termed a "young" or "immature" stream, and the later state a "mature" or "old" stream. However, a stream may meander for some distance before falling into a "young" stream condition.

Intermittent and ephemeral streams

An Australian creek, low in the dry season, carrying little water. The energetic flow of the stream had, in flood, moved finer sediment further downstream. There is a pool to lower right and a riffle to upper left of the photograph.

A perennial stream is one which flows continuously all year. :57 Some perennial streams may only have continuous flow in segments of its stream bed year round during years of normal rainfall. Blue-line streams are perennial streams and are marked on topographic maps with a solid blue line.

Intermittent Stream

In the United States, an intermittent or seasonal stream is one that only flows for part of the year and is marked on topographic maps with a line of blue dashes and dots. A wash or desert wash is normally a dry streambed in the deserts of the American Southwest which flows only after significant rainfall. Washes can fill up quickly during rains, and there may be a sudden torrent of water after a thunderstorm begins upstream, such as during monsoonal conditions. These flash floods often catch travelers by surprise. An intermittent stream can also be called an arroyo in Latin America, a winterbourne in Britain, or a wadi in the Arabic-speaking world.

In Italy an intermittent stream is termed a torrent (Italian torrente). In full flood the stream may or may not be "torrential" in the dramatic sense of the word, but there will be one or more seasons in which the flow is reduced to a trickle or less. Typically torrents have Apennine rather than Alpine sources, and in the summer they are fed by little precipitation and no melting snow. In this case the maximum discharge will be during the spring and autumn. However there are also glacial torrents with a different seasonal regime.

In Australia, an intermittent stream is usually called a creek and marked on topographic maps with a solid blue line.

Ephemeral Stream

Generally, streams that flow only during and immediately after precipitation are termed ephemeral. There is no clear demarcation between surface runoff and an ephemeral stream.

Drainage Basins

The extent of land basin drained by a stream is termed its drainage basin (also known in North America as the watershed and, in British English, as a catchment). A basin may also be composed of smaller basins. For instance, the Continental Divide in North America divides the mainly easterly-draining Atlantic Ocean and Arctic Ocean basins from the largely westerly-flowing Pacific Ocean basin.

The Atlantic Ocean basin, however, may be further subdivided into the Atlantic Ocean and Gulf of Mexico drainages. (This delineation is termed the Eastern Continental Divide.) Similarly, the Gulf of Mexico basin may be divided into the Mississippi River basin and several smaller basins, such as theTombigbee River basin. Continuing in this vein, a component of the Mississippi River basin is the Ohio River basin, which in turn includes theKentucky River basin, and so forth.

5

Description of Hydrologic Cycle

The water, or hydrologic, cycle describes the pilgrimage of water as water molecules make their way from the Earth's surface to the atmosphere and back again, in some cases to below the surface. This gigantic system, powered by energy from the Sun, is a continuous exchange of moisture between the oceans, the atmosphere, and the land.

Studies have revealed that evaporation—the process by which water changes from a liquid to a gas—from oceans, seas, and other bodies of water (lakes, rivers, streams) provides nearly 90% of the moisture in our atmosphere. Most of the remaining 10% found in the atmosphere is released by plants through transpiration. Plants take in water through their roots, then release it through small pores on the underside of their leaves. In addition, a very small portion of water vapor enters the atmosphere through sublimation, the process by which water changes directly from a solid (ice or snow) to a gas. The gradual shrinking of snow banks in cases when the temperature remains below freezing results from sublimation.

Together, evaporation, transpiration, and sublimation, plus volcanic emissions, account for almost all the water vapor in the atmosphere that isn't inserted through human activities. While evaporation from the oceans is the primary vehicle for driving the surface-to-atmosphere portion of the hydrologic cycle, transpiration is also significant. For example, a cornfield 1 acre in size can transpire as much as 4,000 gallons of water every day.

After the water enters the lower atmosphere, rising air currents carry it upward, often high into the atmosphere, where the air is cooler. In the cool air, water vapor is more likely to condense from a gas to a liquid to form cloud droplets. Cloud droplets can grow and produce precipitation (including rain, snow, sleet, freezing rain, and hail), which is the primary mechanism for transporting water from the atmosphere back to the Earth's surface.

When precipitation falls over the land surface, it follows various routes in its subsequent paths. Some of it evaporates, returning to the atmosphere; some seeps into the ground as soil moisture or groundwater; and some runs off into rivers and streams. Almost all of the water eventually flows into the oceans or other bodies of water, where the cycle continues. At different stages of the cycle, some of the water is intercepted by humans or other life forms for drinking, washing, irrigating, and a large variety of other uses.

The hydrologic cycle involves the continuous circulation of water in the Earth-Atmosphere system. At its core, the water cycle is the motion of the water from the ground to the atmosphere and back again. Of the many processes involved in the hydrologic cycle, the most important are... evaporation transpiration condensation precipitation runoff

Evaporation

Evaporation is the change of state in a substance from a liquid to a gas. In meteorology, the substance we are concerned about the most is water.

For evaporation to take place, energy is required. The energy can come from any source: the sun, the atmosphere, the earth, or objects on the earth such as humans.

Everyone has experienced evaporation personally. When the body heats up due to the air temperature or through exercise, the body sweats, secreting water onto the skin.

The purpose is to cause the body to use its heat to evaporate the liquid, thereby removing heat and cooling the body. It is the same effect that can be seen when you step out of a shower or swimming pool. The coolness you feel is from the removing of bodily heat to evaporate the water on your skin.

Transpiration

Transpiration is the evaporation of water from plants through stomata. Stomata are small openings found on the underside of leaves that are connected to vascular plant tissues. In most plants, transpiration is a passive process largely controlled by the humidity of the atmosphere and the moisture content of the soil. Of the transpired water passing through a plant only 1% is used in the growth process of the plant. The remaining 99% is passed into the atmosphere.

Condensation

An example of condensation

Condensation is the process whereby water vapor in the atmosphere is changed into a liquid state. In the atmosphere condensation may appear as clouds or dew. Condensation is the process whereby water appears on the side of an uninsulated cold drink can or bottle.

Condensation is not a matter of one particular temperature but of a difference between two temperatures; the air temperature and the dewpoint temperature. At its basic meaning, the dew point is the temperature where dew can form.

Actually, it is the temperature that, if the air is cool to that level, the air becomes saturated. Any additional cooling causes water vapor to condense. Foggy conditions often occur when air temperature and dew point are equal.

Condensation is the opposite of evaporation. Since water vapor has a higher energy level than that of liquid water, when condensation occurs, the excess energy in the form of heat energy is released. This release of heat aids in the formation of hurricanes.

Earth's Water Cycle

Water is always on the move. Rain falling where you live may have been water in the ocean just days before. And the water you see in a river or stream may have been snow on a high mountaintop.

Water can be in the atmosphere, on the land, in the ocean, and even underground. It is recycled over and over through the water cycle. In the cycle, water changes state between liquid, solid (ice), and gas (water vapor). Most water vapor gets into theatmosphere by a process calledevaporation. This process turns the water that is at the top of the ocean, rivers, and lakes into water vapor in the atmosphere using energy from the Sun. Water vapor can also form from snow and ice through the process of sublimation and can evaporate from plants by a process calledtranspiration.

The water vapor rises in the atmosphereand cools, forming tiny water droplets by a process called condensation. Those water droplets make up clouds. If those tiny water droplets combine with each other they grow larger and eventually become too heavy to stay in the air. Then they fall to the ground as rain, snow, and other types of precipitation

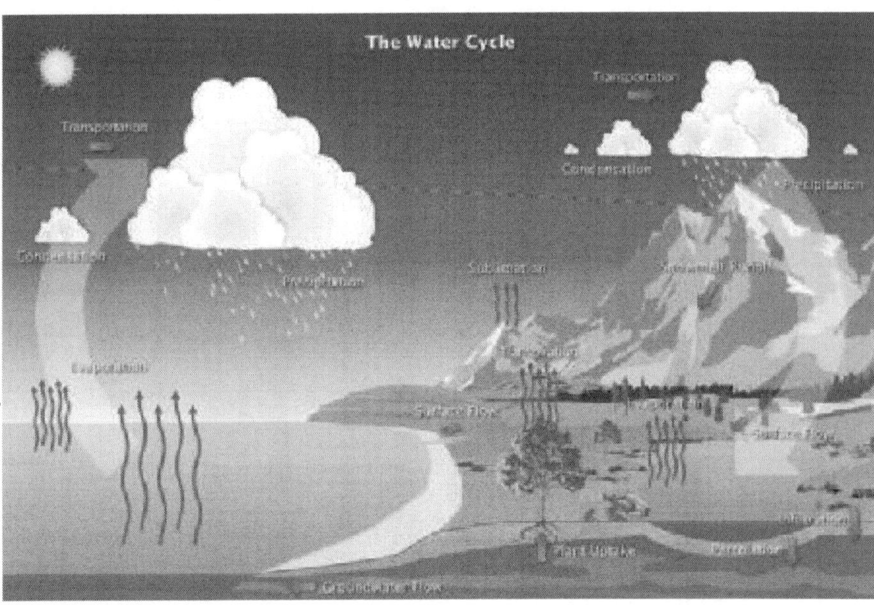

Most of the precipitation that falls becomes a part of the ocean or part of rivers, lakes, and streams that eventually lead to the ocean. Some of the snow and ice that falls as precipitation stays at the Earth surface in glaciers and other types of ice. Some of the precipitation seeps into the ground and becomes a part of the groundwater.

Water stays in certain places longer than others. A drop of water may spend over 3,000 years in the ocean before moving on to another part of the water cycle while a drop of water spends an average of just eight days in the atmosphere before falling back to Earth.

The hydrologic cycle involves the exchange of heat energy, which leads to temperature changes. For instance, in the process of evaporation, water takes up energy from the surroundings and cools the environment. Conversely, in the process of condensation, water releases energy to its surroundings, warming the environment. The water cycle figures significantly in the maintenance of life and ecosystems on Earth.

Even as water in each reservoir plays an important role, the water cycle brings added significance to the presence of water on our planet. By transferring water from one reservoir to another, the water cycle purifies water, replenishes the land with freshwater, and transports minerals to different parts of the globe. It is also involved in reshaping the geological features of the Earth, through such processes as erosion and sedimentation. In addition, as the water cycle also involves heat exchange, it exerts an influence on climate as well.

The sun, which drives the water cycle, heats water in oceans and seas. Water evaporates as water vapor into the air. Ice and snow can sublimate directly into water vapor. Evapotranspiration is water transpired from plants and evaporated from the soil. Rising air currents take the vapor up into the atmosphere where cooler temperatures cause it to condense into clouds. Air currents move water vapor around the globe, cloud particles collide, grow, and fall out of the sky as precipitation. Some precipitation falls as snow or hail, sleet, and can accumulate as ice caps and glaciers, which can store frozen water for thousands of years. Most water falls back into the oceans or onto land as rain, where the water flows over the ground as surface runoff. A portion of runoff enters rivers in valleys in the landscape, with streamflow moving water towards the oceans.

Runoff and groundwater are stored as freshwater in lakes. Not all runoff flows into rivers, much of it soaks into the ground as infiltration. Some water infiltrates deep into the ground and replenishes aquifers, which store freshwater for long periods of time. Some infiltration stays close to the land surface and can seep back into surface-water bodies (and the ocean) as groundwater discharge. Some groundwater finds openings in the land surface and comes out as freshwater springs. Over time, the water returns to the ocean, where our water cycle started.

Changes over Time

The water cycle describes the processes that drive the movement of water throughout the hydrosphere. However, much more water is "in storage" for long periods of time than is actually moving through the cycle. The storehouses for the vast majority of all water on Earth are the oceans. It is estimated that of the 332,500,000 mi3 (1,386,000,000 km3) of the world's water supply, about 321,000,000 mi3 (1,338,000,000 km3) is stored in oceans, or about 95%. It is also estimated that the oceans supply about 90% of the evaporated water that goes into the water cycle.

During colder climatic periods more ice caps and glaciers form, and enough of the global water supply accumulates as ice to lessen the amounts in other parts of the water cycle. The reverse is true during warm periods. During the last ice age glaciers covered almost one-third of Earth's land mass, with the result being that the oceans were about 400 ft (122 m) lower than today. During the last global "warm spell," about 125,000 years ago, the seas were about 18 ft (5.5 m) higher than they are now. About three million years ago the oceans could have been up to 165 ft (50 m) higher.

The scientific consensus expressed in the 2007 Intergovernmental Panel on Climate Change (IPCC) Summary for Policymakers is for the water cycle to continue to intensify throughout the 21st century, though this does not mean that precipitation will increase in all regions. In subtropical land areas — places that are already relatively dry — precipitation is projected to decrease during the 21st century, increasing the probability of drought. The drying is projected to be strongest near the poleward margins of the subtropics (for example, theMediterranean Basin, South Africa, southern Australia, and the Southwestern United States). Annual precipitation amounts are expected to increase in near-equatorial regions that tend to be wet in the present climate, and also at high latitudes. These large-scale patterns are present in nearly all of the climate model simulations conducted at several international research centers as part of the 4th Assessment of the IPCC. There is now ample evidence that increased hydrologic variability and change in climate has and will continue have a profound impact on the water sector through the hydrologic cycle, water availability, water demand, and water allocation at the global, regional, basin, and local levels.

Glacial retreat is also an example of a changing water cycle, where the supply of water to glaciers from precipitation cannot keep up with the loss of water from melting and sublimation.Glacial retreat since 1850 has been extensive.

Human activities that alter the water cycle include:

- agriculture
- industry
- alteration of the chemical composition of the atmosphere
- construction of dams
- deforestation and afforestation
- removal of groundwater from wells
- water abstraction from rivers
- urbanization

Effects on Climate

The water cycle is powered from solar energy. 86% of the global evaporation occurs from the oceans, reducing their temperature by evaporative cooling. Without the cooling, the effect of evaporation on the greenhouse effect would lead to a much higher surface temperature of 67 °C (153 °F), and a warmer planet.

Aquifer drawdown or overdrafting and the pumping of fossil water increases the total amount of water in the hydrosphere that is subject to transpiration and evaporation thereby causingaccretion in water vapour and cloud cover which are the primary absorbers of infrared radiation in the Earth's atmosphere. Adding water to the system has a forcing effect on the whole earth system, an accurate estimate of which hydrogeological fact is yet to be quantified.

Effects on biogeochemical cycling

While the water cycle is itself a biogeochemical cycle, flow of water over and beneath the Earth is a key component of the cycling of other biogeochemicals. Runoff is responsible for almost all of the transport of eroded sediment and phosphorus from land to waterbodies. The salinity of the oceans is derived from erosion and transport of dissolved salts from the land.

Cultural eutrophication of lakes is primarily due to phosphorus, applied in excess to agricultural fields in fertilizers, and then transported overland and down rivers. Both runoff and groundwater flow play significant roles in transporting nitrogen from the land to waterbodies.

The dead zone at the outlet of the Mississippi River is a consequence of nitrates from fertilizer being carried off agricultural fields and funnelled down the river system to the Gulf of Mexico. Runoff also plays a part in the carbon cycle, again through the transport of eroded rock and soil.

Slow Loss over Geologic Time

The hydrodynamic wind within the upper portion of a planet's atmosphere allows light chemical elements such as Hydrogen to move up to the exobase, the lower limit of the exosphere, where the gases can then reach escape velocity, entering outer space without impacting other particles of gas. This type of gas loss from a planet into space is known as planetary wind. Planets with hot lower atmospheres could result in humid upper atmospheres that accelerate the loss of hydrogen.

Bioprecipitation

It is the concept of rain-making bacteria and was proposed by David Sands from Montana State University before 1983. The formation of ice in clouds is required for snowand most rainfall. Dust and soot particles can serve as ice nuclei, but biological ice nuclei are capable of catalyzing freezing at much warmer temperatures. The ice-nucleating bacteria currently known are mostly plant pathogens. Recent research suggests that bacteria may be present in clouds as part of an evolved process of dispersal.

Ice-nucleating Bacteria are Used in Cloud Seeding

Most known ice-nucleating bacteria are plant pathogens. These pathogens can cause freezing injury in plants. In the United States alone, it has been estimated that frost accounts for approximately $1 billion in crop damage each year. The ice-minus variant of P. syringae is a mutant, lacking the gene responsible for ice-nucleating surface protein production. This lack of surface protein provides a less favorable environment for ice formation. Both strains of P. syringae occur naturally, but recombinant DNA technology has allowed for the synthetic removal or alteration of specific genes, enabling the creation of the ice-minus strain.

The introduction of an ice-minus strain of P. syringae to the surface of plants would incur competition between the strains. Should the ice-minus strain win out, the ice nucleate provided by P. syringae would no longer be present, lowering the level of frost development on plant surfaces at normal water freezing temperature (0°C).

Dispersal of Bacteria Through Rainfall

Bacteria present in clouds may have evolved to use rainfall as a means of dispersing themselves. The bacteria are found in snow soils and seedlings in locations such as Antarctica, theYukon Territory of Canada and the French Alps according to Brent Christner, a microbiologist at Louisiana State University. It has been suggested that the bacteria are part of a constant feedback between terrestrial ecosystems and clouds. They may rely on the rainfall to spread to new habitats, in much the same way as plants rely on windblown pollen grains, Christner said, with this possibly a key element of the bacterial life cycle.

Cloud Seeding

Many ski resorts use a commercially available freeze-dried preparation of ice-nucleating bacteria to produce snow as required. This is known as cloud seeding.

Biogeochemical Cycle

In ecology and Earth science, a biogeochemical cycle or substance turnover or cycling of substances is a pathway by which a chemical element or molecule moves through both biotic (biosphere) and abiotic (lithosphere, atmosphere, and hydrosphere) compartments of Earth. A cycle is a series of change which comes back to the starting point and which can be repeated.

The term "biogeochemical" tells us that biological; geological and chemical factors are all involved. On the other hand the circulation of chemical nutrients like carbon, oxygen, nitrogen, phosphorus, calcium, and water etc. through the biological and physical world are known as biogeochemical cycle. In effect, the element is recycled, although in some cycles there may be places (called reservoirs) where the element is accumulated or held for a long period of time (such as an ocean or lake for water).

Water, for example, is always recycled through the water cycle, as shown in the diagram. The water undergoes evaporation, condensation, and precipitation, falling back to Earth clean and fresh. Elements, chemical compounds, and other forms of matter are passed from one organism to another and from one part of the biosphere to another through the biogeochemical cycles.

Ecological systems (ecosystems) have many biogeochemical cycles operating as a part of the system, for example the water cycle, the carbon cycle, the nitrogen cycle, etc. All chemical elements occurring in organisms are part of biogeochemical cycles. In addition to being a part of living organisms, these chemical elements also cycle through abiotic factors of ecosystems such as water (hydrosphere), land (lithosphere), and/or the air (atmosphere).

The living factors of the planet can be referred to collectively as the biosphere. All the nutrients—such as carbon, nitrogen, oxygen, phosphorus, and sulfur—used in ecosystems by living organisms are a part of a closed system; therefore, these chemicals are recycled instead of being lost and replenished constantly such as in an open system.

The flow of energy in an ecosystem is an open system; the sun constantly gives the planet energy in the form of light while it is eventually used and lost in the form of heat throughout the trophic levels of a food web. Carbon is used to make carbohydrates, fats, and proteins, the major sources of food energy. These compounds are oxidized to release carbon dioxide, which can be captured by plants to make organic compounds. The chemical reaction is powered by the light energy of the sun.

It is possible for an ecosystem to obtain energy without sunlight. Carbon must be combined with hydrogen and oxygen in order to be utilized as an energy source, and this process depends on sunlight. Ecosystems in the deep sea, where no sunlight can penetrate, use sulfur. Hydrogen sulfide near hydrothermal vents can be utilized by organisms such as the giant tube worm. In the sulfur cycle, sulfur can be forever recycled as a source of energy. Energy can be released through the oxidation and reduction of sulfur compounds (e.g., oxidizing elemental sulfur to sulfite and then to sulfate).

Although the Earth constantly receives energy from the sun, its chemical composition is essentially fixed, as additional matter is only occasionally added by meteorites. Because this chemical composition is not replenished like energy, all processes that depend on these chemicals must be recycled. These cycles include both the living biosphere and the nonliving lithosphere, atmosphere, and hydrosphere.

Reservoirs

The chemicals are sometimes held for long periods of time in one place. This place is called a reservoir, which, for example, includes such things as coal deposits that are storing carbon for a long period of time. When chemicals are held for only short periods of time, they are being held in exchange pools. Examples of exchange pools include plants and animals.

Plants and animals utilize carbon to produce carbohydrates, fats, and proteins, which can then be used to build their internal structures or to obtain energy. Plants and animals temporarily use carbon in their systems and then release it back into the air or surrounding medium. Generally, reservoirs are abiotic factors whereas exchange pools are biotic factors. Carbon is held for a relatively short time in plants and animals in comparison to coal deposits. The amount of time that a chemical is held in one place is called its residence.

Important Cycles

The most well-known and important biogeochemical cycles, for example, include the carbon cycle, the nitrogen cycle, the oxygen cycle, the phosphorus cycle, the sulfur cycle and the water cycle. There are many biogeochemical cycles that are currently being studied for the first time as climate change and human impacts are drastically changing the speed, intensity, and balance of these relatively unknown cycles. These newly studied biogeochemical cycles include the mercury cycle and the human-caused cycle of atrazine, which may affect certain species.

Biogeochemical cycles always involve hot equilibrium states: a balance in the cycling of the element between compartments. However, overall balance may involve compartments distributed on a global scale.

As biogeochemical cycles describe the movements of substances on the entire globe, the study of these is inherently multidisciplinary. The carbon cycle may be related to research in ecology and atmospheric sciences. Biochemical dynamics would also be related to the fields of geology and pedology (soil study).

Condensation

Condensation is the formation of liquid drops of water from water vapor. It is the process which creates clouds, and so is necessary for rain and snow formation as well. Condensation in the atmosphere usually occurs as a parcel of rising air expands and cools to the point where some of the water vapor molecules clump together faster than they are torn apart from their thermal energy.

A very important part of this process is the release of the latent heat of condensation. This is the heat that was absorbed when the water was originally evaporated from the surface of the Earth, a process which keeps the Earth's surface climate much cooler that it would otherwise be if there were no water.

The heat removed from the surface through evaporation is thereby released again higher up in the atmosphere when clouds form.

Another way in which condensation occurs is on hard surfaces, such as during the formation of dew. Water condensing on a glass of ice water, or on the inside of windows during winter, is the result of those glass surfaces' temperature cooling below the dewpoint of the air which is in contact with them. Condensation is the change of the physical state of matter from gaseous phase into liquid phase, and is the reverse of vaporization. When the transition happens from the gaseous phase into the solid phase directly, the change is called deposition.

Condensation is initiated by the formation of atomic/molecular clusters of that species within its gaseous volume—like rain drop or snow-flake formation within clouds—or at the contact between such gaseous phase and a (solvent) liquid or solid surface.

A few distinct reversibility scenarios emerge here with respect to the nature of the surface.

- absorption into the surface of a liquid (either of the same species or one of its solvents)—is reversible as evaporation.
- adsorption (as dew droplets) onto solid surface at pressures and temperatures higher than the specie's triple point—also reversible as evaporation.
- adsorption onto solid surface (as supplemental layers of solid) at pressures and temperatures lower than the specie's triple point—is reversible as sublimation.

Condensation commonly occurs when a vapour is cooled and/or compressed to its saturation limit when the molecular density in the gas phase reaches its maximal threshold. Vapour cooling and compressing equipment that collects condensed liquids is called "condenser".

Psychrometry measures the rates of condensation from and evaporation into the air moisture at various atmospheric pressures and temperatures. Water is the product of its vapour condensation —condensation is the process of such phase conversion.

Applications of condensation

Condensation is a crucial component of distillation, an important laboratory and industrial chemistry application. Because condensation is a naturally occurring phenomenon, it can often be used to generate water in large quantities for human use. Many structures are made solely for the purpose of collecting water from condensation, such as air wells and fog fences. Such systems can often be used to retain soil moisture in areas where active desertification is occurring—so much so that some organizations educate people living in affected areas about water condensers to help them deal effectively with the situation.

It is also a crucial process in forming particle tracks in a Cloud Chamber. In this case, ions produced by an incident particle act as nucleation centres for the vapour to condense on. Numerous living organisms use water made accessible by condensation. A few examples are the Australian Thorny Devil, Darkling beetles on the Namibian coast and Coast Redwoodson the west coast of the United States.Condensation in building construction is an unwanted phenomenon as it may cause dampness, mold health issues, wood rot, corrosion and energy loss due to increased heat transfer. Interstructure condensation may be caused by thermal bridges, insufficient or lacking damp proofing or insulated glazing.

Evaporation

Evaporation is a type of vaporization of a liquid that occurs only on the surface of a liquid. The other type of vaporization is boiling, which, instead, occurs on the entire mass of the liquid.

On average, the molecules in a glass of water do not have enough heat energy to escape from the liquid. With sufficient heat, the liquid would turn into vapor quickly. When the molecules collide, they transfer energy to each other in varying degrees, based on how they collide. Sometimes the transfer is so one-sided for a molecule near the surface that it ends up with enough energy to 'escape' (evaporate).

Liquids that do not evaporate visibly at a given temperature in a given gas (e.g., cooking oil at room temperature) have molecules that do not tend to transfer energy to each other in a pattern sufficient to frequently give a molecule the heat energy necessary to turn into vapor. However, these liquids are evaporating. It is just that the process is much slower and thus significantly less visible.

Evaporation is an essential part of the water cycle. Solar energy drives evaporation of water from oceans, lakes, moisture in the soil, and other sources of water. In hydrology, evaporation and transpiration (which involves evaporation within plant stomata) are collectively termedevapotranspiration. Evaporation of water occurs when the surface of the liquid is exposed, allowing molecules to escape and form water vapor, this vapor can then rise up and form clouds.

For molecules of a liquid to evaporate, they must be located near the surface, be moving in the proper direction, and have sufficient kinetic energy to overcome liquid-phase intermolecular forces. Only a small proportion of the molecules meet these criteria, so the rate of evaporation is limited. Since the kinetic energy of a molecule is proportional to its temperature, evaporation proceeds more quickly at higher temperatures. As the faster-moving molecules escape, the remaining molecules have lower average kinetic energy, and the temperature of the liquid, thus, decreases. This phenomenon is also called evaporative cooling.

This is why evaporating sweat cools the human body. Evaporation also tends to proceed more quickly with higher flow rates between the gaseous and liquid phase and in liquids with higher vapor pressure. For example, laundry on a clothes line will dry (by evaporation) more rapidly on a windy day than on a still day. Three key parts to evaporation are heat, atmospheric pressure (determines the percent humidity) and air movement.

On a molecular level, there is no strict boundary between the liquid state and the vapor state. Instead, there is a Knudsen layer, where the phase is undetermined. Because this layer is only a few molecules thick, at a macroscopic scale a clear phase transition interface can be seen.

Evaporative Equilibrium

If evaporation takes place in a closed vessel, the escaping molecules accumulate as a vapor above the liquid. Many of the molecules return to the liquid, with returning molecules becoming more frequent as the density and pressure of the vapor increases. When the process of escape and return reaches an equilibrium, the vapor is said to be "saturated," and no further change in either vapor pressure and density or liquid temperature will occur. For a system consisting of vapor and liquid of a pure substance, this equilibrium state is directly related to the vapor pressure of the substance, as given by the Clausius-Clapeyron relation: $\ln\left(\frac{P_2}{P_1}\right) = -\frac{\Delta H_{vap}}{R}\left(\frac{1}{T_2} - \frac{1}{T_1}\right)$

where P1, P2 are the vapor pressures at temperatures T1, T2 respectively, ÄHvap is the enthalpy of vaporization, and R is the universal gas constant. The rate of evaporation in an open system is related to the vapor pressure found in a closed system. If a liquid is heated, when the vapor pressure reaches the ambient pressure the liquid will boil.

The ability for a molecule of a liquid to evaporate is based largely on the amount of kinetic energy an individual particle may possess. Even at lower temperatures, individual molecules of a liquid can evaporate if they have more than the minimum amount of kinetic energy required for vaporization.

6

Applications of GPS in Geographical Studies

Remote sensing is the science of obtaining the physical properties of an area without being there. It allows users to capture, visualize, and analyze objects and features on the Earth's surface. By collecting imagery, we can classify it into land cover and other types of analyses.

The global positioning system (GPS) is a network of satellites and receiving devices used to determine the location of something on Earth. Some GPS receivers are so accurate they can establish their location within 1 centimeter (0.4 inches). GPS receivers provide location in latitude, longitude, and altitude. They also provide the accurate time.

GPS includes 24 satellites that circle Earth in precise orbits. Each satellite makes a full orbit of Earth every 12 hours. These satellites are constantly sending out radio signals. GPS receivers are programmed to receive information about where each satellite is at any given moment. A GPS receiver determines its own location by measuring the time it takes for a signal to arrive at its location from at least four satellites. Because radio waves travel at a constant speed, the receiver can use the time measurements to calculate its distance from each satellite.

Using multiple satellites makes the GPS data more accurate. If a GPS receiver calculates its distance from only one satellite, it could be that exact distance from the satellite in any direction. Think of the satellite as a flashlight. When you shine it on the ground, you get a circle of light. With one satellite, the GPS receiver could be anywhere in that circle of light. With two more satellites, there are two more circles. These three circles intersect, or cross, in only one place. That is the location of the GPS receiver. This method of determining location is called trilateration.

Aircraft, ships, submarines, trains, and the space shuttle all use GPS to navigate. Many people use receivers when driving cars. The GPS receiver plots the car's constantly-changing location on an electronic map. The map provides directions to the person's destination. Both the location and the vehicle are plotted using satellite data. Some hikers use GPS to help them find their way, especially when they are not on marked trails.

Sometimes there are obstacles to getting a clear GPS signal. Gravity can pull the GPS satellites slightly out of orbit. Parts of Earth's atmosphere sometimes distort the satellite radio signals. Trees, buildings, and other structures can also block the radio waves. GPS control and monitoring stations around the world track the satellites and constantly monitor their signals. They then calculate corrections that are broadcast to GPS receivers. These corrections make GPS much more accurate.

The original GPS system began as a project of the U.S. military. The first experimental satellite was launched in 1978. By 1994, a full 24 GPS satellites were orbiting Earth. At first, GPS available for civilian, or nonmilitary, use was not very accurate. It would only locate a GPS receiver within about 300 meters (1,000 feet). Today, an accurate signal is free and available to anyone with a GPS receiver.

Russia has its own version of a GPS system, called GLONASS (Global Orbiting Navigation Satellite System). China and the European Union are currently creating systems of their own.

Geographical involves the study of places and their relationship to people and the environment. It has branches such as human geography, geomatics, physical geography, regional geography, integrated geography, and cartography. Geographical studies seek to know the location of things, reasons they exist, their development, and changes over some time. This makes geographers explore physical features and societies on the earth's surface. For this reason, geographers need GPS to explore the landscape. Keep reading to know various applications of GPS in geographical studies.

1. *Decision making*

It is hard to make the correct decision with the rise of diverse problems. Wrong decisions can cause consequences thus it should be accurate. After geographers have explored different parts of the environment, they use a global positioning system to make decisions. The tool gives highly accurate data that you can use to get the right answers on the occurrence of natural disasters such as earthquakes, landslides, and tsunamis.

2. *Collect accurate data*

Traditionally geographers could not collect efficient information. The global positioning system receiver uses signals from satellites to give accuracy. It collects accurate and efficient data. Besides that, you can use GPS to create dictionaries for collecting data efficiently. This makes it to be a reliable tool for geographers who are carrying out fieldwork.

3. *Monitor environmental changes*

Environmental changes are caused by natural disasters and human activities. These changes include a shortage of freshwater, climatic changes, pollution, mining, deforestation, and loss of biodiversity. The global positioning system has become the main tool for monitoring environmental changes. It gives geographers a warning before the changes. This helps authorities to warn people and government.

4. *Track animals*

In the past years, geographers had one only way to track by following and observing the animals. Also, they could put a tag on the animal until it is recaptured.

Researchers were not sure if the animal will be captured for the second time. Today, geographers use a global positioning system to determine the migration of animals. The GPS is put on the animal's body to track the animal's movement at any time. Furthermore, geographers can know the kind of habitat they use and the size of the range. This important information helps to control the population by allowing reproduction.

5. *Improve travel*

Before the development of the global positioning system, individuals had a hard time when navigating. People had to use physical maps to navigate in a harsh environment. Physical maps are not effective because they can get damaged easily. GPS helps in moving from one unfamiliar place to another hence removes risks of missing. On top of that, it has made travel to be easy.

6. *Improve harvest*

GPS enables the study of soil samples to determine rich-fertile lands. Other applications in the farm are crop scouting, guiding tractor, mapping field, and mapping yields. This land practices help to improve a farmer's harvest.

7. *Determine location*

The device communicates with the satellite to determine a precise position. You will be able to get a distance of reaching your destination.

8. *Map physical features*

Physical features such as mountains, lakes, hills, oceans, seas, deserts, and forests can be put on the map. GPS enables geographers to locate the features accurately.

9. *Gauge effects of erosion*

Soil erosion makes the land to be unfertile. GPS monitors soil erosion of a place.

10. *Prevent natural disasters*

Natural disasters have become a common cause of death today. This technology is installed on the ground to determine natural disasters before they happen. Early warning saves people's lives and property.

Orbital platforms collect and transmit data from different parts of the electromagnetic spectrum, which in conjunction with larger scale aerial or ground-based sensing and analysis, provides researchers with enough information to monitor trends such as El Niño and other natural long and short term phenomena. Other uses include different areas of the earth sciences such as natural resource management, agricultural fields such as land usage and conservation, and national security and overhead, ground-based and stand-off collection on border areas.

By satellite, aircraft, spacecraft, buoy, ship, and helicopter images, data is created to analyze and compare things like vegetation rates, erosion, pollution, forestry, weather, and land use. These things can be mapped, imaged, tracked and observed.

The process of remote sensing is also helpful for city planning, archaeological investigations, military observation and geomorphological surveying.

Data Acquisition Techniques

The basis for multispectral collection and analysis is that of examined areas or objects that reflect or emit radiation that stand out from surrounding areas.

Applications of Remote Sensing Data

- Conventional radar is mostly associated with aerial traffic control, early warning, and certain large scale meteorological data. Doppler radar is used by local law enforcements' monitoring of speed limits and in enhanced meteorological collection such as wind speed and direction within weather systems. Other types of active collection includes plasmas in the ionosphere. Interferometric synthetic aperture radar is used to produce precise digital elevation models of large scale terrain.
- Laser and radar altimeters on satellites have provided a wide range of data. By measuring the bulges of water caused by gravity, they map features on the seafloor to a resolution of a mile or so. By measuring the height and wave-length of ocean waves, the altimeters measure wind speeds and direction, and surface ocean currents and directions.
- Light detection and ranging (LIDAR) is well known in examples of weapon ranging, laser illuminated homing of projectiles. LIDAR is used to detect and measure the concentration of various chemicals in the atmosphere, while airborne LIDAR can be used to measure heights of objects and features on the ground more accurately than with radar technology. Vegetation remote sensing is a principal application of LIDAR.
- Radiometers and photometers are the most common instrument in use, collecting reflected and emitted radiation in a wide range of frequencies. The most common are visible and infrared sensors, followed by microwave, gamma ray and rarely, ultraviolet. They may also be used to detect the emission spectra of various chemicals, providing data on chemical concentrations in the atmosphere.
- Stereographic pairs of aerial photographs have often been used to make topographic maps by imagery and terrain analysts in trafficability and highway departments for potential routes.
- Simultaneous multi-spectral platforms such as Landsat have been in use since the 70's. These thematic mappers take images in multiple wavelengths of electro-magnetic radiation (multi-spectral) and are usually found on Earth observation satellites, including (for example) the Landsat program or the IKONOS satellite. Maps of land cover and land use from thematic mapping can be used to prospect for minerals, detect or monitor land usage, deforestation, and examine the health of indigenous plants and crops, including entire farming regions or forests.
- Hyperspectral imaging produces an image where each pixel has full spectral information with imaging narrow spectral bands over a contiguous spectral range. Hyperspectral imagers are used in various applications including mineralogy, biology, defence, and environmental measurements.

- Within the scope of the combat against desertification, remote sensing allows to follow-up and monitor risk areas in the long term, to determine desertification factors, to support decision-makers in defining relevant measures of environmental management, and to assess their impacts.

Geodetic

- Overhead geodetic collection was first used in aerial submarine detection and gravitational data used in military maps. This data revealed minute perturbations in the Earth'sgravitational field (geodesy) that may be used to determine changes in the mass distribution of the Earth, which in turn may be used for geological studies.

Acoustic and Near-acoustic

- Sonar: passive sonar, listening for the sound made by another object (a vessel, a whale etc); active sonar, emitting pulses of sounds and listening for echoes, used for detecting, ranging and measurements of underwater objects and terrain.
- Seismograms taken at different locations can locate and measure earthquakes (after they occur) by comparing the relative intensity and precise timing.

To coordinate a series of large-scale observations, most sensing systems depend on the following: platform location, what time it is, and the rotation and orientation of the sensor. High-end instruments now often use positional information from satellite navigation systems. The rotation and orientation is often provided within a degree or two with electronic compasses. Compasses can measure not just azimuth (i. e. degrees to magnetic north), but also altitude (degrees above the horizon), since the magnetic field curves into the Earth at different angles at different latitudes. More exact orientations require gyroscopic-aided orientation, periodically realigned by different methods including navigation from stars or known benchmarks.

Resolution impacts collection and is best explained with the following relationship: less resolution=less detail & larger coverage, More resolution=more detail, less coverage. The skilled management of collection results in cost-effective collection and avoid situations such as the use of multiple high resolution data which tends to clog transmission and storage infrastructure.

Data Processing

Generally speaking, remote sensing works on the principle of the inverse problem. While the object or phenomenon of interest (the state) may not be directly measured, there exists some other variable that can be detected and measured (the observation), which may be related to the object of interest through the use of a data-derived computer model. The common analogy given to describe this is trying to determine the type of animal from its footprints. For example, while it is impossible to directly measure temperatures in the upper atmosphere, it is possible to measure the spectral emissions from a known chemical species (such as carbon dioxide) in that region. The frequency of the emission may then be related to the temperature in that region via various thermodynamic relations.

The quality of remote sensing data consists of its spatial, spectral, radiometric and temporal resolutions.

Spatial Resolution

The size of a pixel that is recorded in a raster image – typically pixels may correspond to square areas ranging in side length from 1 to 1,000 metres (3.3 to 3,300 ft).

Spectral Resolution

The wavelength width of the different frequency bands recorded – usually, this is related to the number of frequency bands recorded by the platform. Current Landsat collection is that of seven bands, including several in the infra-red spectrum, ranging from a spectral resolution of 0.07 to 2.1 ìm. The Hyperion sensor on Earth Observing-1 resolves 220 bands from 0.4 to 2.5 ìm, with a spectral resolution of 0.10 to 0.11 ìm per band.

Radiometric Resolution

The number of different intensities of radiation the sensor is able to distinguish. Typically, this ranges from 8 to 14 bits, corresponding to 256 levels of the gray scale and up to 16,384 intensities or "shades" of colour, in each band. It also depends on the instrument noise.

Temporal Resolution

The frequency of flyovers by the satellite or plane, and is only relevant in time-series studies or those requiring an averaged or mosaic image as in deforesting monitoring. This was first used by the intelligence community where repeated coverage revealed changes in infrastructure, the deployment of units or the modification/introduction of equipment. Cloud cover over a given area or object makes it necessary to repeat the collection of said location.

In order to create sensor-based maps, most remote sensing systems expect to extrapolate sensor data in relation to a reference point including distances between known points on the ground. This depends on the type of sensor used. For example, in conventional photographs, distances are accurate in the center of the image, with the distortion of measurements increasing the farther you get from the center. Another factor is that of the platen against which the film is pressed can cause severe errors when photographs are used to measure ground distances. The step in which this problem is resolved is called georeferencing, and involves computer-aided matching up of points in the image (typically 30 or more points per image) which is extrapolated with the use of an established benchmark, "warping" the image to produce accurate spatial data. As of the early 1990s, most satellite images are sold fully georeferenced.

In addition, images may need to be radiometrically and atmospherically corrected.

Radiometric correction gives a scale to the pixel values, e. g. the monochromatic scale of 0 to 255 will be converted to actual radiance values.

Atmospheric correction eliminates atmospheric haze by rescaling each frequency band so that its minimum value (usually realised in water bodies) corresponds to a pixel value of 0. The digitizing of data also make possible to manipulate the data by changing gray-scale values.

Interpretation is the critical process of making sense of the data. The first application was that of aerial photographic collection which used the following process; spatial measurement through the use of a light table in both conventional single or stereographic coverage, added skills such as the use of photogrammetry, the use of photomosaics, repeat coverage, Making use of objects' known dimensions in order to detect modifications. Image Analysis is the recently developed automated computer-aided application which is in increasing use.

Object-Based Image Analysis (OBIA) is a sub-discipline of GIScience devoted to partitioning remote sensing (RS) imagery into meaningful image-objects, and assessing their characteristics through spatial, spectral and temporal scale.

Old data from remote sensing is often valuable because it may provide the only long-term data for a large extent of geography. At the same time, the data is often complex to interpret, and bulky to store. Modern systems tend to store the data digitally, often with lossless compression. The difficulty with this approach is that the data is fragile, the format may be archaic, and the data may be easy to falsify. One of the best systems for archiving data series is as computer-generated machine-readable ultrafiche, usually in typefonts such as OCR-B, or as digitized half-tone images. Ultrafiches survive well in standard libraries, with lifetimes of several centuries. They can be created, copied, filed and retrieved by automated systems. They are about as compact as archival magnetic media, and yet can be read by human beings with minimal, standardized equipment.

Remote Sensing Software

Remote Sensing data is processed and analyzed with computer software, known as a remote sensing application. A large number of proprietary and open source applications exist to process remote sensing data. According to an NOAA Sponsored Research by Global Marketing Insights, Inc. the most used applications among Asian academic groups involved in remote sensing are as follows: ERDAS 36% (ERDAS IMAGINE 25% & ERMapper 11%); ESRI 30%; ITT Visual Information Solutions ENVI 17%; MapInfo 17%. Among Western Academic respondents as follows: ESRI 39%, ERDAS IMAGINE 27%, MapInfo 9%, AutoDesk 7%, ITT Visual Information Solutions ENVI 17%. Other important Remote Sensing Software packages include: TNTmips from MicroImages, PCI Geomatica made by PCI Geomatics, the leading remote sensing software package in Canada, IDRISI from Clark Labs, Image Analyst fromIntergraph, and RemoteView made by Overwatch Textron Systems. Dragon/ips is one of the oldest remote sensing packages still available, and is in some cases free. Open source remote sensing software includes GRASS GIS, ILWIS, QGIS, OSSIM, Opticks (software) and Orfeo toolbox.

Global Positioning System

The Global Positioning System (GPS) is a space-based satellite navigation system that provides location and time information in all weather, anywhere on or near the Earth, where there is an unobstructed line of sight to four or more GPS satellites. It is maintained by the United Statesgovernment and is freely accessible by anyone with a GPS receiver. The system imposes some technical limitations which are only removed for authorized users.

The GPS program provides critical capabilities to military, civil and commercial users around the world. In addition, GPS is the backbone for modernizing the global air traffic system.

The GPS project was developed in 1973 to overcome the limitations of previous navigation systems, integrating ideas from several predecessors, including a number of classified engineering design studies from the 1960s. GPS was created and realized by the U.S. Department of Defense (DoD) and was originally run with 24 satellites. It became fully operational in 1994.

Advances in technology and new demands on the existing system have now led to efforts to modernize the GPS system and implement the next generation of GPS III satellites and Next Generation Operational Control System (OCX). Announcements from the Vice President and the White House in 1998 initiated these changes. In 2000, U.S. Congress authorized the modernization effort, referred to as GPS III.

In addition to GPS, other systems are in use or under development. The Russian GLObal NAvigation Satellite System (GLONASS) was in use by only the Russian military, until it was made fully available to civilians in 2007. There are also the planned European Union Galileo positioning system, Chinese Compass navigation system, and Indian Regional Navigational Satellite System.

The design of GPS is based partly on similar ground-based radio-navigation systems, such as LORAN and the Decca Navigator developed in the early 1940s, and used during World War II. In 1956, Friedwardt Winterberg proposed a test of general relativity (for time slowing in a strong gravitational field) using accurate atomic clocks placed in orbit inside artificial satellites. (To achieve accuracy requirements, GPS uses principles of general relativity to correct the satellites' atomic clocks.) Additional inspiration for GPS came when the Soviet Union launched the first man-made satellite, Sputnik in 1957. Two American physicists, William Guier and George Weiffenbach, at Johns Hopkins's Applied Physics Laboratory (APL), decided on their own to monitor Sputnik's radio transmissions. They soon realized that, because of the Doppler effect, they could pinpoint where the satellite was along its orbit from the Doppler shift. The Director of the APL gave them access to their brand new UNIVAC II to do the heavy calculations required. When they released the orbit of Sputnik to the media, the Russians were dumbfounded to learn how powerful American computers had become, as they would not have been able to calculate the orbit themselves. The following spring, Frank McClure, the deputy director of the APL, asked Guier and Weiffenbach to look at the inverse problem where you know the location of the satellite and you want to find your own location. (The Navy was developing the submarine-launched Polaris missile, which required them to know the submarine's location.) This led them and APL to develop the Transit system.

The first satellite navigation system, Transit (satellite), used by the United States Navy, was first successfully tested in 1960. It used a constellation of five satellites and could provide a navigational fix approximately once per hour. In 1967, the U.S. Navy developed the Timationsatellite that proved the ability to place accurate clocks in space, a technology required by GPS. In the 1970s, the ground-based Omega Navigation System, based on phase comparison of signal transmission from pairs of stations, became the first worldwide radio navigation system. Limitations of these systems drove the need for a more universal navigation solution with greater accuracy.

While there were wide needs for accurate navigation in military and civilian sectors, almost none of those were seen as justification for the billions of dollars it would cost in research, development, deployment, and operation for a constellation of navigation satellites. During the Cold War arms race, the nuclear threat to the existence of the United States was the one need that did justify this cost in the view of the United States Congress. This deterrent effect is why GPS was funded. It is also the reason for the ultra secrecy at that time. The nuclear triadconsisted of the United States Navy's submarine-launched ballistic missiles (SLBMs) along with United States Air Force (USAF) strategic bombers and intercontinental ballistic missiles (ICBMs). Considered vital to the nuclear deterrence posture, accurate determination of the SLBM launch position was a force multiplier.

Precise navigation would enable United States submarines to get an accurate fix of their positions prior to launching their SLBMs. The USAF with two-thirds of the nuclear triad also had requirements for a more accurate and reliable navigation system. The Navy and Air Force were developing their own technologies in parallel to solve what was essentially the same problem. To increase the survivability of ICBMs, there was a proposal to use mobile launch platforms (such as Russian SS-24 and SS-25) and so the need to fix the launch position had similarity to the SLBM situation.

In 1960, the Air Force proposed a radio-navigation system called MOSAIC (MObile System for Accurate ICBM Control) that was essentially a 3-D LORAN. A follow-on study called Project 57 was worked in 1963 and it was "in this study that the GPS concept was born." That same year the concept was pursued as Project 621B, which had "many of the attributes that you now see in GPS" and promised increased accuracy for Air Force bombers as well as ICBMs. Updates from the Navy Transit system were too slow for the high speeds of Air Force operation. The Navy Research Laboratory continued advancements with their Timation (Time Navigation) satellites, first launched in 1967, and with the third one in 1974 carrying the first atomic clock into orbit.

With these parallel developments in the 1960s, it was realized that a superior system could be developed by synthesizing the best technologies from 621B, Transit, Timation, and SECOR in a multi-service program.

During Labor Day weekend in 1973, a meeting of about 12 military officers at the Pentagon discussed the creation of a Defense Navigation Satellite System (DNSS). It was at this meeting that "the real synthesis that became GPS was created." Later that year, the DNSS program was namedNavstar. With the individual satellites being associated with the name Navstar (as with the predecessors Transit and Timation), a more fully encompassing name was used to identify the constellation of Navstar satellites, Navstar-GPS, which was later shortened simply to GPS.

After Korean Air Lines Flight 007, carrying 269 people, was shot down in 1983 after straying into the USSR's prohibited airspace, in the vicinity ofSakhalin and Moneron Islands, President Ronald Reagan issued a directive making GPS freely available for civilian use, once it was sufficiently developed, as a common good. The first satellite was launched in 1989, and the 24th satellite was launched in 1994.

Initially, the highest quality signal was reserved for military use, and the signal available for civilian use was intentionally degraded (Selective Availability). This changed with President Bill Clinton ordering Selective Availability to be turned off at midnight May 1, 2000, improving the precision of civilian GPS from 100 meters (330 ft) to 20 meters (66 ft). The executive order signed in 1996 to turn off Selective Availability in 2000 was proposed by the US Secretary of Defense, William Perry, because of the widespread growth of differential GPS services to improve civilian accuracy and eliminate the US military advantage. Moreover, the US military was actively developing technologies to deny GPS service to potential adversaries on a regional basis.

Over the last decade, the U.S. has implemented several improvements to the GPS service, including new signals for civil use and increased accuracy and integrity for all users, all while maintaining compatibility with existing GPS equipment.

GPS modernization has now become an ongoing initiative to upgrade the Global Positioning System with new capabilities to meet growing military, civil, and commercial needs. The program is being implemented through a series of satellite acquisitions, including GPS Block III and the Next Generation Operational Control System (OCX). The U.S. Government continues to improve the GPS space and ground segments to increase performance and accuracy.

GPS is owned and operated by the United States Government as a national resource. Department of Defense (DoD) is the steward of GPS. Interagency GPS Executive Board (IGEB)oversaw GPS policy matters from 1996 to 2004. After that the National Space-Based Positioning, Navigation and Timing Executive Committee was established by presidential directive in 2004 to advise and coordinate federal departments and agencies on matters concerning the GPS and related systems. The executive committee is chaired jointly by the deputy secretaries of defense and transportation. Its membership includes equivalent-level officials from the departments of state, commerce, and homeland security, the joint chiefs of staff, and NASA. Components of the executive office of the president participate as observers to the executive committee, and the FCC chairman participates as a liaison.

The DoD is required by law to "maintain a Standard Positioning Service (as defined in the federal radio navigation plan and the standard positioning service signal specification) that will be available on a continuous, worldwide basis," and "develop measures to prevent hostile use of GPS and its augmentations without unduly disrupting or degrading civilian uses."

Basic Concept of GPS

A GPS receiver calculates its position by precisely timing the signals sent by GPS satellites high above the Earth. Each satellite continually transmits messages that include

- the time the message was transmitted
- precise orbital information (the ephemeris)
- the general system health and rough orbits of all GPS satellites (the almanac).

The receiver uses the messages it receives to determine the transit time of each message and computes the distance to each satellite. These distances along with the satellites' locations are used with the possible aid of trilateration, depending on which algorithm is used, to compute the position of the receiver. This position is then displayed, perhaps with a moving map display or latitude and longitude; elevation information may be included. Many GPS units show derived information such as direction and speed, calculated from position changes.

Three satellites might seem enough to solve for position since space has three dimensions and a position near the Earth's surface can be assumed. However, even a very small clock error multiplied by the very large speed of light — the speed at which satellite signals propagate — results in a large positional error. Therefore receivers use four or more satellites to solve for both the receiver's location and time. The very accurately computed time is effectively hidden by most GPS applications, which use only the location. A few specialized GPS applications do however use the time; these include time transfer, traffic signal timing, and synchronization of cell phone base stations.

Although four satellites are required for normal operation, fewer apply in special cases. If one variable is already known, a receiver can determine its position using only three satellites. For example, a ship or aircraft may have known elevation. Some GPS receivers may use additional clues or assumptions (such as reusing the last known altitude, dead reckoning,inertial navigation, or including information from the vehicle computer) to give a less accurate (degraded) position when fewer than four satellites are visible.

To provide an introductory description of how a GPS receiver works, error effects are deferred to a later section. Using messages received from a minimum of four visible satellites, a GPS receiver is able to determine the times sent and then the satellite positions corresponding to these times sent. The x, y, and z components of position, and the time sent, are designated as $[x_i, y_i, z_i, t_i]$ where the subscript i is the satellite number and has the value 1, 2, 3, or 4. Knowing the indicated time the message was received $\bar{t}_r$, the GPS receiver could compute the transit time of the message as $\left(\bar{t}_r - t_i\right)$, if would be equal to correct reception time, . A pseudorange, , would be the traveling distance of the message, assuming it traveled at the speed of light, c.

A satellite's position and pseudorange define a sphere, centered on the satellite, with radius equal to the pseudorange. The position of the receiver is somewhere on the surface of this sphere. Thus with four satellites, the indicated position of the GPS receiver is at or near the intersection of the surfaces of four spheres. In the ideal case of no errors, the GPS receiver would be at a precise intersection of the four surfaces.

If the surfaces of two spheres intersect at more than one point, they intersect in a circle. The article trilateration shows this mathematically. A figure, Two Sphere Surfaces Intersecting in a Circle, is shown below. Two points where the surfaces of the spheres intersect are clearly shown in the figure. The distance between these two points is the diameter of the circle of intersection. The intersection of a third spherical surface with the first two will be its intersection with that circle; in most cases of practical interest, this means they intersect at two points. Another figure, Surface of Sphere Intersecting a Circle (not a solid disk) at Two Points, illustrates the intersection. The two intersections are marked with dots. Again the article trilateration clearly shows this mathematically.

For automobiles and other near-earth vehicles, the correct position of the GPS receiver is the intersection closest to the Earth's surface. For space vehicles, the intersection farthest from Earth may be the correct one.

The correct position for the GPS receiver is also the intersection closest to the surface of the sphere corresponding to the fourth satellite.

Correcting a GPS Receiver's Clock

One of the most significant error sources is the GPS receiver's clock. Because of the very large value of the speed of light, c, the estimated distances from the GPS receiver to the satellites, the pseudoranges, are very sensitive to errors in the GPS receiver clock; for example an error of one microsecond (0.000 001 second) corresponds to an error of 300 metres (980 ft). This suggests that an extremely accurate and expensive clock is required for the GPS receiver to work. Because manufacturers prefer to build inexpensive GPS receivers for mass markets, the solution for this dilemma is based on the way sphere surfaces intersect in the GPS problem.

It is likely that the surfaces of the three spheres intersect, because the circle of intersection of the first two spheres is normally quite large, and thus the third sphere surface is likely to intersect this large circle. It is very unlikely that the surface of the sphere corresponding to the fourth satellite will intersect either of the two points of intersection of the first three, because any clock error could cause it to miss intersecting a point. However, the distance from the valid estimate of GPS receiver position to the surface of the sphere corresponding to the fourth satellite can be used to compute a clock correction.

Let , which is the distance from the valid estimate of GPS receiver position to the fourth satellite, and let denote the pseudorange of the fourth satellite. Let , which is the distance from the computed GPS receiver position to the surface of the sphere corresponding to the fourth satellite. Thus the quotient, , provides an estimate of GPS receiver's clock bias: , where is the time indicated by the receiver's on-board clock and is the correct reception time. The GPS receiver clock can be advanced if is positive or delayed if is negative.

Structure

The current GPS consists of three major segments. These are the space segment (SS), a control segment (CS), and a user segment (US). The U.S. Air Force develops, maintains, and operates the space and control segments. GPS satellites broadcast signals from space, and each GPS receiver uses these signals to calculate its three-dimensional location (latitude, longitude, and altitude) and the current time.

The space segment is composed of 24 to 32 satellites in medium Earth orbit and also includes the payload adapters to the boosters required to launch them into orbit. The control segment is composed of a master control station, an alternate master control station, and a host of dedicated and shared ground antennas and monitor stations. The user segment is composed of hundreds of thousands of U.S. and allied military users of the secure GPS Precise Positioning Service, and tens of millions of civil, commercial, and scientific users of the Standard Positioning Service.

Control Segment

The control segment is composed of

1. a master control station (MCS),
2. an alternate master control station,
3. four dedicated ground antennas and
4. six dedicated monitor stations

The MCS can also access U.S. Air Force Satellite Control Network (AFSCN) ground antennas (for additional command and control capability) and NGA (National Geospatial-Intelligence Agency) monitor stations. The flight paths of the satellites are tracked by dedicated U.S. Air Force monitoring stations in Hawaii, Kwajalein, Ascension Island, Diego Garcia, Colorado Springs, Colorado and Cape Canaveral, along with shared NGA monitor stations operated in England, Argentina, Ecuador, Bahrain, Australia and Washington DC. The tracking information is sent to the Air Force Space Command's MCS at Schriever Air Force Base 25 km (16 mi) ESE of Colorado Springs, which is operated by the 2nd Space Operations Squadron (2 SOPS) of the U.S. Air Force. Then 2 SOPS contacts each GPS satellite regularly with a navigational update using dedicated or shared (AFSCN) ground antennas (GPS dedicated ground antennas are located at Kwajalein, Ascension Island, Diego Garcia, and Cape Canaveral). These updates synchronize the atomic clocks on board the satellites to within a few nanoseconds of each other, and adjust the ephemeris of each satellite's internal orbital model. The updates are created by a Kalman filter that uses inputs from the ground monitoring stations, space weather information, and various other inputs.

Satellite maneuvers are not precise by GPS standards. So to change the orbit of a satellite, the satellite must be marked unhealthy, so receivers will not use it in their calculation. Then the maneuver can be carried out, and the resulting orbit tracked from the ground. Then the new ephemeris is uploaded and the satellite marked healthy again.

The Operation Control Segment (OCS) currently serves as the control segment of record. It provides the operational capability that supports global GPS users and keeps the GPS system operational and performing within specification.

OCS successfully replaced the legacy 1970's-era mainframe computer at Schriever Air Force Base in September 2007. After installation, the system helped enable upgrades and provide a foundation for a new security architecture that supported the U.S. armed forces. OCS will continue to be the ground control system of record until the new segment, Next Generation GPS Operation Control System (OCX), is fully developed and functional.

The new capabilities provided by OCX will be the cornerstone for revolutionizing GPS's mission capabilities, and enabling Air Force Space Command to greatly enhance GPS operational services to U.S. combat forces, civil partners and myriad of domestic and international users.

The GPS OCX program also will reduce cost, schedule and technical risk. It is designed to provide 50% sustainment cost savings through efficient software architecture and Performance-Based Logistics. In addition, GPS OCX expected to cost millions less than the cost to upgrade OCS while providing four times the capability.

The GPS OCX program represents a critical part of GPS modernization and provides significant information assurance improvements over the current GPS OCS program.

- OCX will have the ability to control and manage GPS legacy satellites as well as the next generation of GPS III satellites, while enabling the full array of military signals.
- Built on a flexible architecture that can rapidly adapt to the changing needs of today's and future GPS users allowing immediate access to GPS data and constellations status through secure, accurate and reliable information.

§ Empowers the warfighter with more secure, actionable and predictive information to enhance situational awareness.

- Enables new modernized signals (L1C, L2C, and L5) and has M-code capability, which the legacy system is unable to do.
- Provides significant information assurance improvements over the current program including detecting and preventing cyber attacks, while isolating, containing and operating during such attacks.
- Supports higher volume near real-time command and control capability.

On September 14, 2011, the U.S. Air Force announced the completion of GPS OCX Preliminary Design Review and confirmed that the OCX program is ready for the next phase of development.

The GPS OCX program has achieved major milestones and is on track to support the GPS IIIA launch in May 2014.

User Segment

The user segment is composed of hundreds of thousands of U.S. and allied military users of the secure GPS Precise Positioning Service, and tens of millions of civil, commercial and scientific users of the Standard Positioning Service. In general, GPS receivers are composed of an antenna, tuned to the frequencies transmitted by the satellites, receiver-processors, and a highly stable clock (often a crystal oscillator). They may also include a display for providing location and speed information to the user. A receiver is often described by its number of channels: this signifies how many satellites it can monitor simultaneously. Originally limited to four or five, this has progressively increased over the years so that, as of 2007, receivers typically have between 12 and 20 channels.

GPS receivers may include an input for differential corrections, using the RTCM SC-104 format. This is typically in the form of an RS-232 port at 4,800 bit/s speed. Data is actually sent at a much lower rate, which limits the accuracy of the signal sent using RTCM . Receivers with internal DGPS receivers can outperform those using external RTCM data . As of 2006, even low-cost units commonly include Wide Area Augmentation System (WAAS) receivers.

Many GPS receivers can relay position data to a PC or other device using the NMEA 0183 protocol. Although this protocol is officially defined by the National Marine Electronics Association (NMEA), references to this protocol have been compiled from public records, allowing open source tools like gpsd to read the protocol without violating intellectual property laws. Other proprietary protocols exist as well, such as the SiRF and MTK protocols. Receivers can interface with other devices using methods including a serial connection, USB, or Bluetooth.

Applications

While originally a military project, GPS is considered a dual-use technology, meaning it has significant military and civilian applications.

GPS has become a widely deployed and useful tool for commerce, scientific uses, tracking, and surveillance. GPS's accurate time facilitates everyday activities such as banking, mobile phone operations, and even the control of power grids by allowing well synchronized hand-off switching.

Civilian

Many civilian applications use one or more of GPS's three basic components: absolute location, relative movement, and time transfer.

- Clock synchronization: The accuracy of GPS time signals (±10 ns) is second only to the atomic clocks upon which they are based.
- Cellular telephony: Clock synchronization enables time transfer, which is critical for synchronizing its spreading codes with other base stations to facilitate inter-cell handoff and support hybrid GPS/cellular position detection for mobile emergency calls and other applications. The first handsets with integrated GPS launched in the late 1990s. The U.S. Federal Communications Commission (FCC) mandated the feature in either the handset or in the towers (for use in

triangulation) in 2002 so emergency services could locate 911 callers. Third-party software developers later gained access to GPS APIs from Nextel upon launch, followed by Sprint in 2006, and Verizon soon thereafter.

- Disaster relief/emergency services: Depend upon GPS for location and timing capabilities.
- Geofencing: Vehicle tracking systems, person tracking systems, and pet tracking systems use GPS to locate a vehicle, person, or pet. These devices are attached to the vehicle, person, or the pet collar. The application provides continuous tracking and mobile or Internet updates should the target leave a designated area.
- Geotagging: Applying location coordinates to digital objects such as photographs and other documents for purposes such as creating map overlays.
- GPS Aircraft Tracking
- GPS tours: Location determines what content to display; for instance, information about an approaching point of interest.
- Map-making: Both civilian and military cartographers use GPS extensively.
- Navigation: Navigators value digitally precise velocity and orientation measurements.
- Phasor measurements: GPS enables highly accurate timestamping of power system measurements, making it possible to computephasors.
- Robotics: Self-navigating, autonomous robots using a GPS sensors, which calculate latitude, longitude, time, speed, and heading.
- Recreation: For example, geocaching, geodashing, GPS drawing and waymarking.
- Surveying: Surveyors use absolute locations to make maps and determine property boundaries.
- Tectonics: GPS enables direct fault motion measurement in earthquakes.
- Telematics: GPS technology integrated with computers and mobile communications technology in automotive navigation systems
- Fleet Tracking: The use of GPS technology to identify, locate and maintain contact reports with one or more fleet vehicles in real-time.

Communication

The navigational signals transmitted by GPS satellites encode a variety of information including satellite positions, the state of the internal clocks, and the health of the network. These signals are transmitted on two separate carrier frequencies that are common to all satellites in the network. Two different encodings are used: a public encoding that enables lower resolution navigation, and an encrypted encoding used by the U.S. military.

Demodulation and Decoding

Because all of the satellite signals are modulated onto the same L1 carrier frequency, the signals must be separated after demodulation. This is done by assigning each satellite a unique binary sequence known as a Gold code. The signals are decoded after demodulation using addition of the Gold codes corresponding to the satellites monitored by the receiver.

If the almanac information has previously been acquired, the receiver picks the satellites to listen for by their PRNs, unique numbers in the range 1 through 32. If the almanac information is not in memory, the receiver enters a search mode until a lock is obtained on one of the satellites. To obtain a lock, it is necessary that there be an unobstructed line of sight from the receiver to the satellite. The receiver can then acquire the almanac and determine the satellites it should listen for. As it detects each satellite's signal, it identifies it by its distinct C/A code pattern. There can be a delay of up to 30 seconds before the first estimate of position because of the need to read the ephemeris data.

Processing of the navigation message enables the determination of the time of transmission and the satellite position at this time. For more information see Demodulation and Decoding, Advanced.

Navigation Equations

The receiver uses messages received from satellites to determine the satellite positions and time sent. The x, y, and z components of satellite position and the time sent are designated as xi, yi, zi, ti] where the subscript i denotes the satellite and has the value 1, 2, ..., n, where $n \geq 4$. Knowing when the message was received t_r, the receiver computes the message's transit time as $t_r - t_i$. Note that the receiver indeed knows the reception time indicated by its on-board clock, $\tilde{t}_r$ rather than . Assuming the message traveled at the speed of light (c) the distance traveled is (tr " ti)c. Knowing the distance from receiver to satellite and the satellite's position implies that the receiver is on the surface of a sphere centered at the satellite's position. Thus the receiver is at or near the intersection of the surfaces of the spheres. In the ideal case of no errors, the receiver is at the intersection of the surfaces of the spheres.

Let b denote the clock error or bias, the amount that the receiver's clock is off. The receiver has four unknowns, the three components of GPS receiver position and the clock bias [x, y, z, b]. The equations of the sphere surfaces are given by: $(x - x_i)^2 + (y - y_i)^2 + (z - z_i)^2 \quad ([\tilde{t}_r + b - t_i]c)^2, \; i \quad 1, 2, \ldots, n$

or in terms of pseudoranges, $p_i = (\tilde{t}_r - t_i)\,c$, as

$$p_i = \sqrt{(x - x_i)^2 + (y - y_i)^2 + (z - z_i)^2} - bc, \; i = 1, 2, \ldots, n$$

These equations can be solved by algebraic or numerical methods.

Bancroft's Method

Bancroft's method involves an algebraic as opposed to numerical method and can be used for the case of four satellites or for the case of more than four satellites. If there are four satellites then Bancroft's method provides the unique solution for the four unknowns. If there are more than four satellites then Bancroft's method provides the solution which minimizes the sum of the squares of the errors for the over determined system.

Trilateration

The receiver can use trilateration and one dimensional numerical root finding. Trilateration is used to determine the position based on three satellite's pseudoranges. In the usual case of two intersections, the point nearest the surface of the sphere corresponding to the fourth satellite is chosen. Let d denote the signed distance from the receiver position to the sphere around the fourth satellite. The notation, d(correction) shows this as a function of the correction, because it changes the pseudoranges. The problem is to determine the correction such that d(correction) = 0. This is the familiar problem of finding the zeroes of a one dimensional non-linear function of a scalar variable. Iterative numerical methods, such as those found in the chapter on root finding in Numerical Recipes can solve this type of problem.

Additional Methods for More than four Satellites

When more than four satellites are available, the calculation can use the four best or more than four, considering number of channels, processing capability, and geometric dilution of precision (GDOP). Using more than four is an over-determined system of equations with no unique solution, which must be solved by least-squares or a similar technique. If all visible satellites are used, the results are as good as or better than using the four best. Errors can be estimated through the residuals. With each combination of four or more satellites, a GDOP factor can be calculated, based on the relative sky directions of the satellites used.

As more satellites are picked up, pseudoranges from various 4-way combinations can be processed to add more estimates to the location and clock offset. The receiver then takes the weighted average of these positions and clock offsets. After the final location and time are calculated, the location is expressed in a specific coordinate system such as latitude and longitude, using the WGS 84 geodetic datum or a country-specific system.

Error sources and Analysis

Error analysis for the Global Positioning System is an important aspect for determining what errors and their magnitude are to be expected. GPS errors are affected by geometric dilution of precision and depend on signal arrival time errors, numerical errors, atmospherics effects, ephemeris errors, multipath errors and other effects. The single largest source of error in modelling the orbital dynamics is due to variability in solar radiation pressure.

Augmentation

Integrating external information into the calculation process can materially improve accuracy. Such augmentation systems are generally named or described based on how the information arrives. Some systems transmit additional error information (such as clock drift, ephemera, or ionospheric delay), others characterize prior errors, while a third group provides additional navigational or vehicle information.

Examples of augmentation systems include the Wide Area Augmentation System (WAAS), European Geostationary Navigation Overlay Service (EGNOS), Differential GPS, Inertial Navigation Systems (INS) and Assisted GPS.

Precise Monitoring

Accuracy can be improved through precise monitoring and measurement of existing GPS signals in additional or alternate ways.

The largest remaining error is usually the unpredictable delay through the ionosphere. The spacecraft broadcast ionospheric model parameters, but errors remain. This is one reason GPS spacecraft transmit on at least two frequencies, L1 and L2. Ionospheric delay is a well-defined function of frequency and the total electron content (TEC) along the path, so measuring the arrival time difference between the frequencies determines TEC and thus the precise ionospheric delay at each frequency.

Military receivers can decode the P(Y)-code transmitted on both L1 and L2. Without decryption keys, it is still possible to use a codeless technique to compare the P(Y) codes on L1 and L2 to gain much of the same error information. However, this technique is slow, so it is currently available only on specialized surveying equipment. In the future, additional civilian codes are expected to be transmitted on the L2 and L5 frequencies. Then all users will be able to perform dual-frequency measurements and directly compute ionospheric delay errors.

Relative Kinematic Positioning (RKP) is a third alternative for a precise GPS-based positioning system. In this approach, determination of range signal can be resolved to a precision of less than 10 centimetres (3.9 in). This is done by resolving the number of cycles that the signal is transmitted and received by the receiver by using a combination of differential GPS (DGPS) correction data, transmitting GPS signal phase information and ambiguity resolution techniques via statistical tests—possibly with processing in real-time (real-time kinematic positioning, RTK).

Timekeeping and Leap Seconds

While most clocks are synchronized to Coordinated Universal Time (UTC), the atomic clocks on the satellites are set to GPS time (GPST; see the page of United States Naval Observatory). The difference is that GPS time is not corrected to match the rotation of the Earth, so it does not contain leap seconds or other corrections that are periodically added to UTC. GPS time was set to match Coordinated Universal Time (UTC) in 1980, but has since diverged. The lack of corrections means that GPS time remains at a constant offset withInternational Atomic Time (TAI) (TAI – GPS = 19 seconds). Periodic corrections are performed on the on-board clocks to keep them synchronized with ground clocks.

The GPS navigation message includes the difference between GPS time and UTC, which as of 2011 is 15 seconds because of the leap second added to UTC December 31, 2008. Receivers add this offset to GPS time to calculate UTC and specific timezone values. New GPS units may not show the correct UTC time until after receiving the UTC offset message. The GPS-UTC offset field can accommodate 255 leap seconds (eight bits) that, given the current period of the Earth's rotation (with one leap second introduced approximately every 18 months), should be sufficient to last until approximately the year 2300.

7

Composition and Structure of the Earth's Atmosphere

We all know that earth is a unique planet due to the presence of life. The air is one among the necessary conditions for the existence of life on this planet. The air is a mixture of several gases and it encompasses the earth from all sides. The air surrounding the earth is called the atmosphere.

Atmospheric processes and energy exchanges are driven by Earth's energy balance and linked to climate and weather. Hurricanes, thunderstorms, tornadoes, blizzards, ice storms, dust storms, heat waves, as well as flash flooding resulting from intense precipitation, are all natural processes that are hazardous to people. These severe hazards affect considerable portions of Northern America and are responsible for causing significant death and destruction each year. But in order to understand topics such as thunderstorms, tornadoes, lightning, and hurricanes, we need to spend some time looking at the basics of weather.

Incoming Solar Radiation

The Sun is the driving force of our weather and climate. Its diameter is about 865,000 miles or 109 Earths lined up side by side and is so large that it makes up 99.99 percent of the mass of our Solar System. Our star - the Sun - is the most common type of star in the universe and has existed for nearly 4.6 billion years.

The composition of the Sun is about 92 percent hydrogen (the lightest element in the universe) and almost 8 percent helium (the second lightest element). Using these elements, the Sun generates energy through a process called thermonuclear reaction. This energy is created when the Sun fuses together hydrogen atoms to make a heavier element called helium. This process not only creates new elements, but also releases an enormous amount of energy.

Composition and structure of the atmosphere

Atmosphere is the air surrounding the earth.

The atmosphere is a mixture of different gases. It contains life-giving gases like Oxygen for humans and animals and carbon dioxide for plants.

It envelops the earth all round and is held in place by the gravity of the earth.

It helps in stopping the ultraviolet rays harmful to the life and maintains the suitable temperature necessary for life.

Generally, atmosphere extends up to about 1600 km from the earth's surface. However, 99 % of the total mass of the atmosphere is confined to the height of 32 km from the earth's surface.

Composition of the atmosphere

The atmosphere is made up of different gases, water vapour and dust particles.

The composition of the atmosphere is not static and it changes according to the time and place.

Air is mainly composed of nitrogen, oxygen, and argon, which together constitute the major gases of the atmosphere. The remaining gases are often referred to as trace gases, among which are the greenhouse gases such as water vapor, carbon dioxide, methane, nitrous oxide, and ozone. Filtered air includes trace amounts of many other chemical compounds. Many natural substances may be present in tiny amounts in an unfiltered air sample, including dust, pollen and spores, sea spray, and volcanic ash. Various industrial pollutants also may be present, such as chlorine (elementary or in compounds), fluorine compounds, elemental mercury, and sulfur compounds such as sulfur dioxide [SO2].

The energy emitted by the Sun is called electromagnetic energy, which travels at the speed of light in a spectrum of various wavelengths. (Recall from the module on tsunamis that a wavelength is the distance between two wave crests and frequency is the number of wavelengths that pass a given point in a given amount of time.)

The sun's life expectancy is already half over and in another 4 billion years it will consume all the hydrogen within it leaving behind helium and a few other newly created elements. When the hydrogen runs out, in about 5 billion years, the sun will grow into a red giant, expanding its diameter well past Mercury and possibly Venus and Earth. The sun will continue to fuse helium into carbon until it begins to release its outer layers to form a planetary nebula. Left behind will be the white hot core of our star, called a white dwarf, that will slowly fade out over billions of years.

Warming Up the Atmosphere

Why would you want to wear a black shirt in the winter and a white shirt in the summer? Or would you rather walk barefoot on black pavement or grass? This relates to the idea that different objects absorb and reflect energy differently on the planet. Dark objects tend to absorb more energy, whereas lighter objects reflect more energy. The percent of energy an object reflects is an object's albedo. Because the planet has so many different albedos (cities, forests, deserts, oceans, rocks, glaciers, etc.), the planet has uneven heating on the planet. The result is that of all the energy the Earth receives from the Sun, only 51 percent is actually absorbed by the planet while 49 percent is reflected back into space.

Uneven heating of the planet also occurs because of the curvature of the Earth. The planet receives most of its energy near the equator where the Sun's rays are directly (or nearly directly) hitting the planet. As you move toward the poles, the Sun's energy becomes more diffused and spread out over a greater area. Ultimately, the equator receives more energy from the Sun than it can radiate back into space. There is a surplus of energy. At the poles, the planet radiates more energy out into space than it receives from the Sun. There is a deficit of energy in these regions. The ultimate purpose of weather is to transfer the surplus of energy and heat to the poles and bring colder air toward the equator and find equilibrium.

Global Transfer of Heat

Picture

If the planet is trying to balance the extremes of the poles and equator, how is heat transferred around the planet? There are four ways: conduction, convection, latent heat, and radiation.

CONDUCTION

Conduction is the transfer of heat from a warmer object to a cooler object through molecule interaction. As the Sun heats the ground, energy is transferred to the atmosphere by conduction. But the atmosphere is a very poor conductor of heat. In calm weather, the heated ground only warms the first few centimeters of air. In fact, the atmosphere's temperature can be 50 percent colder at your head than near your feet. Since the atmosphere is such a poor conductor of heat, there must be other ways to transfer the energy.

CONVECTION

Convection is the transfer of heat by the mass movement of a fluid (such as water and air). It occurs mostly in liquids and gases because they are free to move around. Heat is transferred upward and outward away from its heat source and cooler air is brought in to replace the rising air. On a local scale convection in the summer can produce afternoon thunderstorms, but on a global scale convection transfers energy from the equator toward the poles and back.

RADIATION

The final process of transferring heat around the planet is called radiation. The energy coming from the Sun comes in various wavelengths. A wavelength is the distance measured along a wave of energy from one crest to another; the shorter the wavelength, the higher the energy. The energy received from the Sun passes through the atmosphere without warming it. But all objects that absorb radiation from the sun radiate some of that energy back out in a weaker form of energy called long wave radiation such as heat. All living objects (including humans) radiate longwave radiation. The Earth itself radiates the energy absorbed from the Sun in the form of longwave radiation which is sometimes called Earth light. If you were an astronaut wearing infrared goggles, you would actually see the earth glowing from the radiating infrared radiation.

LATENT HEAT

Latent heat, also called a phase change, is a powerful force in weather. When water transforms from gas to liquid or solid (or vise versa), it is called a phase change. The heat required to change phases is called latent heat. In order for water to change from a liquid to a gas, energy/heat must be taken from some external source such as the surrounding atmosphere. Therefore, evaporation is a cooling process because the water is taking heat from the surrounding air in order to evaporate. With the image above, visualize that every time evaporation or condensation occurs, there is a direct transfer of heat and energy.

The process of evaporation and condensation transfers large amounts of heat around the planet. The heat released in a typical thunderstorm is equal to an atomic bomb. The heat/energy within a hurricane is equal to 100,000 atomic bombs and could power the United States for over a year!

Latent heat, also called a phase change, is a powerful force in weather. When water transforms from gas to liquid or solid (or vise versa), it is called a phase change. The heat required to change phases is called latent heat. In order for water to change from a liquid to a gas, energy/heat must be taken from some external source such as the surrounding atmosphere. Therefore, evaporation is a cooling process because the water is taking heat from the surrounding air in order to evaporate. With the image above, visualize that every time evaporation or condensation occurs, there is a direct transfer of heat and energy.

The process of evaporation and condensation transfers large amounts of heat around the planet. The heat released in a typical thunderstorm is equal to an atomic bomb. The heat/energy within a hurricane is equal to 100,000 atomic bombs and could power the United States for over a year!

Structure of The Atmosphere

In general, air pressure and density decrease in the atmosphere as height increases. However, temperature has a more complicated profile with altitude. Because the general pattern of this profile is constant and recognizable through means such as balloon soundings, temperature provides a useful metric to distinguish between atmospheric layers. In this way, Earth's atmosphere can be divided into five main layers. From highest to lowest, these layers are:

Exosphere

An exosphere is the uppermost layer of an atmosphere where the density is extremely low. An upward traveling molecule moving through the exosphere fast enough to attain escape velocity can escape to space with a low chance of collisions; if it is moving below escape velocity it will be prevented from escaping from the celestial body by gravity. In either case, such a molecule is unlikely to collide with another molecule due to the exosphere's low density.

The main gases within the Earth's exosphere are the lightest gases, mainly hydrogen, with some helium, carbon dioxide, and atomic oxygen near the exobase. The exosphere is the last layer before outer space. Since there is no clear boundary between outer space and the exosphere, the exosphere is sometimes considered a part of outer space.

Lower Boundary

The altitude of its lower boundary, known as the thermopause and exobase, ranges from about 250 to 500 kilometres (160 to 310 mi) depending on solar activity. Its lower boundary at the edge of the thermosphere has sometimes been estimated to be 500 to 1,000 km (310 to 620 mi) above the Earth's surface. The exobase is also called the critical level, the lowest altitude of the exosphere, and is typically defined in one of two ways:

1. The height above which there are negligible atomic collisions between the particles (free molecular flow) and
2. The height above which constituent atoms are on purely ballistic trajectories.

If we define the exobase as the height at which upward traveling molecules experience one collision on average, then at this position the mean free path of a molecule is equal to one pressure scale height. This is shown in the following. Consider a volume of air, with horizontal area A and height equal to the mean free path l, at pressure p and temperature T. For an ideal gas, the number of molecules contained in it is:

$$n = \frac{pAl}{RT}$$

where R is the universal gas constant. From the requirement that each molecule traveling upward undergoes on average one collision, the pressure is: $p = \frac{m_A n g}{A}$

where mA is the mean molecular mass of the gas. Solving these two equations gives: $l = \frac{RT}{m_A g}$

which is the equation for the pressure scale height. As the pressure scale height is almost equal to the density scale height of the primary constituent, and since the Knudsen number is the ratio of mean free path and typical density fluctuation scale, this means that the exobase lies in the region where $\mathrm{Kn}(h_{EB}) \simeq 1$.

The fluctuation in the height of the exobase is important because this provides atmospheric drag on satellites, eventually causing them to fall from orbit if no action is taken to maintain the orbit.

Upper Boundary

The upper boundary of the exosphere can be defined theoretically by the altitude about 190,000 kilometres (120,000 mi), half the distance to the Moon, at which the influence of solar radiation pressure on atomic hydrogen velocities exceeds that of the Earth's gravitational pull. The exosphere observable from space as thegeocorona is seen to extend to at least 100,000 kilometres (62,000 mi) from the surface of the

Earth. The exosphere is a transitional zone between Earth's atmosphere and interplanetary space.

The atmosphere has changed a lot compared to the Earth's early atmosphere, but for the last billion years it has remained pretty constant. We now need to look at 3 very different atmospheric problems:

(1) The Greenhouse effect : The earth is surrounded by a blanket of gases. This blanket traps energy in the atmosphere, much the same way as glass traps heat inside a greenhouse. This results in an build up of energy, and the overall warming of the atmosphere. The greenhouse effect is a natural process which made life on Earth possible. Without naturally occurring greenhouse gases such as water vapour, carbon dioxide, methane and nitrous oxide, the Earth's surface temperature would be 33°C cooler, a chilly -18°C rather than the tolerable 15°C.

When we talk about the greenhouse effect we mean the ENHANCED effect which is caused by the increase of greenhouse gases from human sources. Since the beginning of industrialization, 200 years ago, concentrations of these gases have increased . It is estimated that the Earth's average temperature has risen by 0.6°C since 1880 because of emissions of greenhouse gases from human activity.

The main sources of these emissions, particularly carbon dioxide, methane and nitrous oxide, are:

- The combustion of large amounts of fossil fuels (producing CO2)
- Deforestation (less trees mean that less CO2 is being mopped up)

A increase in global temperatures may seem great, you might even think of 'Costa del Blackpool'. Unfortunately global warming will probably result in big swings in weather patterns across the world. Summers will become dryer and hotter, Winters will be wetter and colder. Other things will start to happen:

- Thermal expansion of the water and melting of continental glaciers would cause sea levels to rise, possibly as much as two feet, by the end of next century.
- Rising temperatures could lead to changes in regional wind systems which would influence global rainfall distribution and lead to the redistribution and frequency of floods, droughts and forest fires.
- Increased sea temperatures would cause the destruction of coral reefs around the world.
- Climate change would create favourable conditions for growth in insect populations. This would likely have a bad effect on agriculture and human health and result in a spread of malaria and other tropical diseases.
- Water supplies would become disrupted and droughts would be more common.

There is a lot of controversy surrounding global warming, views range from those who believe that there is nothing to worry about to those who believe that the world is heading for a global catastrophe. An edited version of a Greenpeace article on Global warning, climate change and the greenhouse effect can be found (here). Other sources of information can be found on the links page on this site

(2) Damage to the ozone layer : Ozone is oxygen that contains molecules that have 3 oxygen atoms (O3). The molecule is triatomic instead of the usual O2 molecule which is diatomic. There is a layer of ozone high up in the atmosphere which shields the Earth from the sun's harmful UV rays, these rays can lead to an increase in skin cancer. The ozone is present in very small quantities but it is enough to absorb the UV rays preventing them reaching the surface.

Scientists began to investigate the ozone layer in the 1970's, it wasn't until the mid 1980's that alarm bells started to ring. Concentrations of ozone appeared to be dropping in certain areas of the world (the layer was starting to thin-out). The cause of this reduction was thought to be man-made.

In 1985 over 60 countries pledged to phased out a group of chemicals called CFC's. These very stable chemicals were once widely used in aerosols and refrigerators. It was thought that their release into the atmosphere produced chlorine radicals which reacted with O3 to produce O2. The emission of CFC's into the environment is now greatly reduced, unfortunately the damage has already been done and the CFC molecules, thanks to their stability, are still causing ozone depletion.

(3) Acid rain : Rain water is naturally acidic due to carbon dioxide which partially reacts with water to give carbonic acid (H2O + CO2 -> H2CO3). When we talk about acid rain we mean the ENHANCED effect which is caused by other gases released when fossil fuels are burnt. Two gases are the main culprits:

Sulphur dioxide - Fossil fuels often contain a lot of sulphur impurities which burn to give sulphur dioxide. The SO2 reacts with water in the atmosphere to from a weak solution of sulphuric acid.

Nitrogen oxides - Under normal conditions nitrogen and oxygen don't react together. At very high temperatures (in an engine) a small proportion of oxygen reacts with nitrogen to givenitrogen oxides. These oxides react with water in the atmosphere to from a weak solution of nitric acid acid.

The dilute acid falls to ground as acid rain which causes the following problems:

- Lakes become acidic and plants and fishes die as a result
- Tree growth is damaged, whole forests can die as a result
- Acid rain attacks metal structures and also buildings made of limestone

Thermosphere

The thermosphere is the biggest of all the layers of the Earth's atmosphere directly above the mesosphere and directly below the exosphere. Within this layer,ultraviolet radiation causes ionization. The International Space Station has a stable orbit within the middle of the thermosphere, between 320 and 380 kilometres (200 and 240 mi). Auroras also occur in the thermosphere.

Named from the Greek thermos meaning heat, the thermosphere begins about 80 kilometres (50 mi) above the Earth. At these high altitudes, the residual atmospheric gases sort into strata according to molecular mass. Thermospheric temperatures increase with altitude due to absorption of highly energetic solar radiation by the small amount of residual oxygen still present. Temperatures are highly dependent on solar activity, and can rise to 1,500 °C (2,730 °F). Radiation causes the atmosphere particles in this layer to become electrically charged, enabling radio waves to bounce off and be received beyond the horizon. In the exosphere, beginning at 500 to 1,000 kilometres (310 to 620 mi) above the Earth's surface, the atmosphere turns into space.

The highly diluted gas in this layer can reach 2,500 °C (4,530 °F) during the day. Even though the temperature is so high, one would not feel warm in the thermosphere, because it is so near vacuum that there is not enough contact with the few atoms of gas to transfer much heat. A normal thermometer would read significantly below 0 °C (32 °F), due to the energy lost by thermal radiation overtaking the energy acquired from the atmospheric gas by direct contact. In theanacoustic zone above 160 kilometres (99 mi), the density is so low that molecular interactions are too infrequent to permit the transmission of sound.

The dynamics of the lower thermosphere (below approximately 120 kilometres (75 mi)) are dominated by atmospheric tide, which is driven, in part, by the very significant diurnal heating. The atmospheric tide dissipates above this level since molecular concentrations do not support the coherent motion needed for fluid flow.

Mesosphere

The mesosphere is the layer of the Earth's atmosphere that is directly above the stratosphere and directly below the thermosphere. In the mesosphere temperature decreases with increasing height. The upper boundary of the mesosphere is the mesopause, which can be the coldest naturally occurring place on Earth with temperatures below 130 K. The exact upper and lower boundaries of the mesosphere vary with latitude and with season, but the lower boundary of the mesosphere is usually located at heights of about 50 km above the Earth's surface and the mesopause is usually at heights near 100 km, except at middle and high latitudes in summer where it descends to heights of about 85 km.

The stratosphere, mesosphere and lowest part of the thermosphere are collectively referred to as the "middle atmosphere", which spans heights from approximately 10 to 100 km. The mesopause, at an altitude of 80–90 km (50–56 mi), separates the mesosphere from the thermosphere—the second-outermost layer of the Earth's atmosphere. This is also around the same altitude as the turbopause, below which different chemical species are well mixed due to turbulent eddies. Above this level the atmosphere becomes non-uniform; the scale heights of different chemical species differ by their molecular masses.

Temperature

Within the mesosphere, temperature decreases with increasing altitude. This is due to decreasing solar heating and increasing cooling by CO2 radiative emission. The top of the mesosphere, called the mesopause, is the coldest part of Earth's atmosphere. Temperatures in the upper mesosphere fall as low as "100 °C(173 K; "148 °F), varying according to latitude and season.

Dynamical Features

The main dynamical features in this region are strong zonal (East-West) winds, atmospheric tides, internal atmospheric gravity waves (commonly called "gravity waves") and planetary waves. Most of these tides and waves are excited in the troposphere and lower stratosphere, and propagate upward to the mesosphere. In the mesosphere, gravity-wave amplitudes can become so large that the waves become unstable and dissipate. This dissipation deposits momentum into the mesosphere and largely drives global circulation.

Noctilucent clouds are located in the mesosphere. The upper mesosphere is also the region of the ionosphere known as the D layer. The D layer is only present during the day, when some ionization occurs with nitric oxide being ionized by Lyman series-alpha hydrogen radiation. The ionization is so weak that when night falls, and the source of ionization is removed, the free electron and ion form back into a neutral molecule. A 5 km (3.1 mi) deep sodium layer is located between 80–105 km (50–65 mi). Made of unbound, non-ionized atoms of sodium, the sodium layer radiates weakly to contribute to the airglow.

Uncertainties

The mesosphere lies above the maximum altitude for aircraft and below the minimum altitude for orbital spacecraft. It has only been accessed through the use ofsounding rockets. As a result, it is the most poorly understood part of the atmosphere. The presence of red sprites and blue jets (electrical discharges or lightningwithin the lower mesosphere), noctilucent clouds and density shears within the poorly understood layer are of current scientific interest.

Meteors

Millions of meteors enter the atmosphere, an average of 40 tons per day. Within the mesosphere most melt or vaporize as a result of collisions with the gas particles contained there. This results in a higher concentration of iron and other refractory materials reaching the surface.

Stratosphere

The stratosphere is the second major layer of Earth's atmosphere, just above the troposphere, and below the mesosphere. It is stratifiedin temperature, with warmer layers higher up and cooler layers farther down. This is in contrast to the troposphere near the Earth's surface, which is cooler higher up and warmer farther down. The border of the troposphere and stratosphere, the tropopause, is marked by where this inversion begins, which in terms of atmospheric thermodynamics is the equilibrium level. The stratosphere is situated between about 10 km (6 mi) and 50 km (30 mi) altitude above the surface at moderate latitudes, while at the poles it starts at about 8 km (5 mi) altitude.

Ozone and Temperature

Within this layer, temperature increases as altitude increases; the top of the stratosphere has a temperature of about 270 K ("3°C or 29.6°F), just slightly below the freezing point of water. The stratosphere is layered in temperature because ozone (O_3) here absorbs high energy UVB and UVC energy waves from the Sun and is broken down into atomic oxygen (O) and diatomic oxygen (O_2).

Atomic oxygen is found prevalent in the upper stratosphere due to the bombardment of UV light and the destruction of both ozone and diatomic oxygen. The mid stratosphere has less UV light passing through it, O and O_2 are able to combine, and is where the majority of natural ozone is produced. It is when these two forms of oxygen recombine to form ozone that they release the heat found in the stratosphere.

The lower stratosphere receives very low amounts of UVC, thus atomic oxygen is not found here and ozone is not formed (with heat as the byproduct). This vertical stratification, with warmer layers above and cooler layers below, makes the stratosphere dynamically stable: there is no regular convection and associated turbulence in this part of the atmosphere. The top of the stratosphere is called the stratopause, above which the temperature decreases with height.

Methane, (CH_4) while not a direct cause of ozone destruction in the stratosphere, does lead to the formation of compounds that destroy ozone. Monoatomic oxygen (O) in the upper stratosphere reacts with methane (CH_4) to form a hydroxyl radical (OH·). This hydroxyl radical is then able to interact with non-soluble compounds like chlorofluorocarbons, and UV light break off chlorine radicals (Cl·). These chlorine radicals break off an oxygen atom from the ozone molecule, creating an oxygen molecule (O_2) and a hypochlorite radical (ClO·). The hypochlorite radical then reacts with an atomic oxygen creating another oxygen molecule and another chlorine radical, thereby preventing the reaction of a monoatomic oxygen with O_2 to create natural ozone.

Aircraft Flight

Commercial airliners typically cruise at altitudes of 9–12 km (30,000–39,000 ft) in temperate latitudes (in the lower reaches of the stratosphere). This optimizes fuel burn, mostly thanks to the low temperatures encountered near the tropopause and low air density, reducing parasitic drag on the airframe. It also allows them to stay above hard weather (extreme turbulence).

Because the temperature in the tropopause and lower stratosphere remains constant (or slightly increases) with increasing altitude, very little convective turbulence occurs at these altitudes. Though most turbulence at this altitude is caused by variations in the jet stream and other local wind shears, areas of significant convective activity (thunderstorms) in the troposphere below may produce convective overshoot.

Although a few gliders have achieved great altitudes in the powerful thermals in thunderstorms, this is dangerous. Most high altitude flights by gliders use lee wavesfrom mountain ranges and were used to set the current record of 15,447 m (50,679 ft).

Circulation and Mixing

The stratosphere is a region of intense interactions among radiative, dynamical, and chemical processes, in which the horizontal mixing of gaseous components proceeds much more rapidly than in vertical mixing.

An interesting feature of stratospheric circulation is the quasi-biennial oscillation (QBO) in the tropical latitudes, which is driven by gravity waves that are convectively generated in the troposphere. The QBO induces a secondary circulation that is important for the global stratospheric transport of tracers, such asozone or water vapor.

In northern hemispheric winter, sudden stratospheric warmings, caused by the absorption of Rossby waves in the stratosphere, can often be observed.

Life

Bacterial life survives in the stratosphere, making it a part of the biosphere. Also, some bird species have been reported to fly at the lower levels of the stratosphere. On November 29, 1975, a Rüppell's Vulture was reportedly ingested into a jet engine 11,552 m (37,900 ft) above the Ivory Coast, and Bar-headed geese routinely overfly Mount Everest's summit, which is 8,848 m (29,029 ft).

Troposphere

The troposphere is the lowest portion of Earth's atmosphere. It contains approximately 80% of the atmosphere's mass and 99% of its water vapor and aerosols. The average depth of the troposphere is approximately 17 km (11 mi) in the middle latitudes. It is deeper in the tropics, up to 20 km (12 mi), and shallower near thepolar regions, at 7 km (4.3 mi) in summer, and indistinct in winter.

The lowest part of the troposphere, where friction with the Earth's surface influences air flow, is the planetary boundary layer. This layer is typically a few hundred meters to 2 km (1.2 mi) deep depending on the landform and time of day. The border between the troposphere and stratosphere, called the tropopause, is a temperature inversion.

The word troposphere derives from the Greek: tropos for "turning" or "mixing," reflecting the fact that turbulent mixing plays an important role in the troposphere's structure and behavior. Most of the phenomena we associate with day-to-day weather occur in the troposphere.

Composition

The chemical composition of the troposphere is essentially uniform, with the notable exception of water vapor. The source of water vapor is at the surface through the processes of evaporation and transpiration. Furthermore the temperature of the troposphere decreases with height, and saturation vapor pressure decreases strongly as temperature drops, so the amount of water vapor that can exist in the atmosphere decreases strongly with height. Thus the proportion of water vapor is normally greatest near the surface and decreases with height.

Pressure

The pressure of the atmosphere is maximum at sea level and decreases with higher altitude. This is because the atmosphere is very nearly in hydrostatic equilibrium, so that the pressure is equal to the weight of air above a given point. The change in pressure with height, therefore can be equated to the density with this hydrostatic equation: $\frac{dp}{dz} = -\rho g_n = -\frac{mpg}{RT}$

where:

- gn is the standard gravity
- r is the density
- z is the altitude
- p is the pressure
- R is the gas constant
- T is the thermodynamic (absolute) temperature
- m is the molar mass

Since temperature in principle also depends on altitude, one needs a second equation to determine the pressure as a function of height, as discussed in the next section.

Temperature

The temperature of the troposphere generally decreases as altitude increases. The rate at which the temperature decreases, " dT / dz, is called the environmental lapse rate (ELR). The ELR is nothing more than the difference in temperature between the surface and the tropopause divided by the height. The reason for this temperature difference is the absorption of the sun's energy occurs at the ground which heats the lower levels of the atmosphere, and the radiation of heat occurs at the top of the atmosphere cooling the earth, this process maintaining the overall heat balance of the earth.

As parcels of air in the atmosphere rise and fall, they also undergo changes in temperature for reasons described below. The rate of change of the temperature in the parcel may be less than or more than the ELR. When a parcel of air rises, it expands, because the pressure is lower at higher altitudes. As the air parcel expands, it pushes on the air around it, doing work; but generally it does not gain heat in exchange from its environment, because its thermal conductivity is low (such a process is called adiabatic). Since the parcel does work and gains no heat, it loses energy, and so its temperature decreases. (The reverse, of course, will be true for a sinking parcel of air.)

Since the heat exchanged dQ is related to the entropy change dS by dQ = TdS, the equation governing the temperature as a function of height for a thoroughly mixed atmosphere is $\frac{dS}{dz} = 0$

where S is the entropy. The rate at which temperature decreases with height under such conditions is called the adiabatic lapse rate.

For dry air, which is approximately an ideal gas, we can proceed further. The adiabatic equation for an ideal gas is $p(z)T(z)^{-\frac{\gamma}{\gamma-1}} = constant$

where ã is the heat capacity ratio (ã=7/5, for air). Combining with the equation for the pressure, one arrives at the dry adiabatic lapse rate,

$$\frac{dT}{dz} = -\frac{mg}{R}\frac{\gamma - 1}{\gamma} = -9.8°\mathrm{C/km}$$

If the air contains water vapor, then cooling of the air can cause the water to condense, and the behavior is no longer that of an ideal gas. If the air is at the saturated vapor pressure, then the rate at which temperature drops with height is called the saturated adiabatic lapse rate. More generally, the actual rate at which the temperature drops with altitude is called the environmental lapse rate. In the troposphere, the average environmental lapse rate is a drop of about 6.5 °C for every 1 km (1,000 meters) in increased height.

The environmental lapse rate (the actual rate at which temperature drops with height, dT / dz) is not usually equal to the adiabatic lapse rate (or correspondingly, $dS/dz \neq 0$). If the upper air is warmer than predicted by the adiabatic lapse rate (dS / dz > 0), then when a parcel of air rises and expands, it will arrive at the new height at a lower temperature than its surroundings. In this case, the air parcel is denser than its surroundings, so it sinks back to its original height, and the air is stable against being lifted. If, on the contrary, the upper air is cooler than predicted by the adiabatic lapse rate, then when the air parcel rises to its new height it will have a higher temperature and a lower density than its surroundings, and will continue to accelerate upward.

Temperatures decrease at middle latitudes from an average of 15°C at sea level to about -55°C at the top of the tropopause. At the poles, the troposphere is thinner and the temperature only decreases to -45°C, while at the equator the temperature at the top of the troposphere can reach -75°C.

Tropopause

The tropopause is the boundary region between the troposphere and the stratosphere. Measuring the temperature change with height through the troposphere and the stratosphere identifies the location of the tropopause. In the troposphere, temperature decreases with altitude. In the stratosphere, however, the temperature remains constant for a while and then increases with altitude. The region of the atmosphere where the lapse rate changes from positive (in the troposphere) to negative (in the stratosphere), is defined as the tropopause. Thus, the tropopause is an inversion layer, and there is little mixing between the two layers of the atmosphere.

Atmospheric Flow

The flow of the atmosphere generally moves in a west to east direction. This however can often become interrupted, creating a more north to south or south to north flow. These scenarios are often described in meteorology as zonal or meridional.

These terms, however, tend to be used in reference to localised areas of atmosphere (at a synoptic scale)). A fuller explanation of the flow of atmosphere around the Earth as a whole can be found in the three-cell model.

Zonal Flow

A zonal flow regime is the meteorological term meaning that the general flow pattern is west to east along the Earth's latitude lines, with weak shortwaves embedded in the flow. The use of the word "zone" refers to the flow being along the Earth's latitudinal "zones". This pattern can buckle and thus become a meridional flow.

Meridional Flow

When the zonal flow buckles, the atmosphere can flow in a more longitudinal (or meridional) direction, and thus the term "meridional flow" arises. Meridional flow patterns feature strong, amplified troughs and ridges, with more north-south flow in the general pattern than west-to-east flow.

Three-cell Model

The three cells model attempts to describe the actual flow of the Earth's atmosphere as a whole. It divides the Earth into the tropical (Hadley cell), mid latititude (Ferrel cell), and polar (polar cell) regions, dealing with energy flow and global circulation. Its fundamental principle is that of balance - the energy that the Earth absorbs from the sun each year is equal to that which it loses back into space, but this however is not a balance precisely maintained in each latitude due to the varying strength of the sun in each "cell" resulting from the tilt of the Earth's axis in relation to its orbit. It demonstrates that a pattern emerges to mirror that of the ocean - the tropics do not continue to get warmer because the atmosphere transports warm air poleward and cold air equatorward, the purpose of which appears to be that of heat and moisture distribution around the planet.

Synoptic scale observations and concepts

Forcing

Forcing is a term used by meteorologists to describe the situation where a change or an event in one part of the atmosphere causes a strengthening change in another part of the atmosphere. It is usually used to describe connections between upper, middle or lower levels (such as upper-level divergence causing lower level convergence in cyclone formation), but can sometimes also be used to describe such connections over distance rather than height alone. In some respects, tele-connections could be considered a type of forcing.

Divergence and Convergence

An area of convergence is one in which the total mass of air is increasing with time, resulting in an increase in pressure at locations below the convergence level (recall that atmospheric pressure is just the total weight of air above a given point).

Divergence is the opposite of convergence - an area where the total mass of air is decreasing with time, resulting in falling pressure in regions below the area of divergence.

Where divergence is occurring in the upper atmosphere, there will be air coming in to try to balance the net loss of mass (this is called the principle of mass conservation), and there is a resulting upward motion (positive vertical velocity). Another way to state this is to say that regions of upper air divergence are conducive to lower level convergence, cyclone formation, and positive vertical velocity. Therefore, identifying regions of upper air divergence is an important step in forecasting the formation of a surface low pressure area.

Other Layers

Within the five principal layers determined by temperature are several layers determined by other properties:

- The ozone layer is contained within the stratosphere. In this layer ozone concentrations are about 2 to 8 parts per million, which is much higher than in the lower atmosphere but still very small compared to the main components of the atmosphere. It is mainly located in the lower portion of the stratosphere from about 15–35 km (9.3–22 mi; 49,000–110,000 ft), though the thickness varies seasonally and geographically. About 90% of the ozone in our atmosphere is contained in the stratosphere.
- The ionosphere, the part of the atmosphere that is ionized by solar radiation, stretches from 50 to 1,000 km (31 to 620 mi; 160,000 to 3,300,000 ft) and typically overlaps both the exosphere and the thermosphere. It forms the inner edge of the magnetosphere. It has practical importance because it influences, for example, radio propagation on the Earth. It is responsible for auroras.
- The homosphere and heterosphere are defined by whether the atmospheric gases are well mixed. In the homosphere the chemical composition of the atmosphere does not depend on molecular weight because the gases are mixed by turbulence. The homosphere includes the troposphere, stratosphere, and mesosphere. Above the turbopause at about 100 km (62 mi; 330,000 ft) (essentially corresponding to the mesopause), the composition varies with altitude. This is because the distance that particles can move without colliding with one another is large compared with the size of motions that cause mixing. This allows the gases to stratify by molecular weight, with the heavier ones such as oxygen and nitrogen present only near the bottom of the heterosphere. The upper part of the heterosphere is composed almost completely of hydrogen, the lightest element.
- The planetary boundary layer is the part of the troposphere that is nearest the Earth's surface and is directly affected by it, mainly through turbulent diffusion. During the day the planetary boundary layer usually is well-mixed, while at night it becomes stably stratified with weak or intermittent mixing. The depth of the planetary boundary layer ranges from as little as about 100 m on clear, calm nights to 3000 m or more during the afternoon in dry regions.

The average temperature of the atmosphere at the surface of Earth is 14 °C (57 °F; 287 K) or 15 °C (59 °F; 288 K), depending on the reference.

Physical Properties

Pressure and Thickness

The average atmospheric pressure at sea level is about 1 atmosphere (atm) = 101.3 kPa (kilopascals) = 14.7 psi (pounds per square inch) = 760 torr = 29.92 inches of mercury (symbol Hg). Total atmospheric mass is 5.1480×1018 kg (1.135×1019 lb), about 2.5% less than would be inferred from the average sea level pressure and the Earth's area of 51007.2 megahectares, this portion being displaced by the Earth's mountainous terrain. Atmospheric pressure is the total weight of the air above unit area at the point where the pressure is measured. Thus air pressure varies with location and weather.

If atmospheric density were to remain constant with height the atmosphere would terminate abruptly at 8.50 km (27,900 ft). Instead, density decreases with height, dropping by 50% at an altitude of about 5.6 km (18,000 ft). As a result the pressure decrease is approximately exponential with height, so that pressure decreases by a factor of two approximately every 5.6 km (18,000 ft) and by a factor of e = 2.718... approximately every 7.64 km (25,100 ft), the latter being the average scale height of Earth's atmosphere below 70 km (43 mi; 230,000 ft).

However, because of changes in temperature, average molecular weight, and gravity throughout the atmospheric column, the dependence of atmospheric pressure on altitude is modeled by separate equations for each of the layers listed above. Even in the exosphere, the atmosphere is still present. This can be seen by the effects of atmospheric drag on satellites.

In summary, the equations of pressure by altitude in the above references can be used directly to estimate atmospheric thickness. However, the following published data are given for reference:

- 50% of the atmosphere by mass is below an altitude of 5.6 km (18,000 ft).
- 90% of the atmosphere by mass is below an altitude of 16 km (52,000 ft). The common altitude of commercial airliners is about 10 km (33,000 ft) and Mt. Everest's summit is 8,848 m (29,029 ft) above sea level.
- 99.99997% of the atmosphere by mass is below 100 km (62 mi; 30,000 ft), although in the rarefied region above this there are auroras and other atmospheric effects. The highest X-15 plane flight in 1963 reached an altitude of 108.0 km (354,300 ft).

Density and Mass

The density of air at sea level is about 1.2 kg/m3 (1.2 g/L). Density is not measured directly but is calculated from measurements of temperature, pressure and humidity using the equation of state for air (a form of the ideal gas law). Atmospheric density decreases as the altitude increases. This variation can be approximately modeled using the barometric formula. More sophisticated models are used to predict orbital decay of satellites.

The average mass of the atmosphere is about 5 quadrillion (5×1015) tonnes or 1/1,200,000 the mass of Earth. According to the AmericanNational Center for

Atmospheric Research, "The total mean mass of the atmosphere is 5.1480×1018 kg with an annual range due to water vapor of 1.2 or 1.5×1015 kg depending on whether surface pressure or water vapor data are used; somewhat smaller than the previous estimate. The mean mass of water vapor is estimated as 1.27×1016 kg and the dry air mass as 5.1352 ±0.0003×1018 kg."

Optical Properties

Solar radiation (or sunlight) is the energy the Earth receives from the Sun. The Earth also emits radiation back into space, but at longer wavelengths that we cannot see. Part of the incoming and emitted radiation is absorbed or reflected by the atmosphere.

Scattering

When light passes through our atmosphere, photons interact with it through scattering. If the light does not interact with the atmosphere, it is called direct radiation and is what you see if you were to look directly at the Sun. Indirect radiation is light that has been scattered in the atmosphere. For example, on an overcast day when you cannot see your shadow there is no direct radiation reaching you, it has all been scattered.

As another example, due to a phenomenon called Rayleigh scattering, shorter (blue) wavelengths scatter more easily than longer (red) wavelengths. This is why the sky looks blue, you are seeing scattered blue light. This is also why sunsets are red. Because the Sun is close to the horizon, the Sun's rays pass through more atmosphere than normal to reach your eye. Much of the blue light has been scattered out, leaving the red light in a sunset.

Absorption

Different molecules absorb different wavelengths of radiation. For example, O_2 and O_3 absorb almost all wavelengths shorter than 300 nanometers. Water (I I2O) absorbs many wavelengths above 700 nm. When a molecule absorbs a photon, it increases the energy of the molecule. We can think of this as heating the atmosphere, but the atmosphere also cools by emitting radiation, as discussed below.

The combined absorption spectra of the gases in the atmosphere leave "windows" of low opacity, allowing the transmission of only certain bands of light. The optical window runs from around 300 nm (ultraviolet-C) up into the range humans can see, the visible spectrum (commonly called light), at roughly 400–700 nm and continues to the infrared to around 1100 nm. There are also infrared and radio windows that transmit some infrared and radio waves at longer wavelengths. For example, the radio window runs from about one centimeter to about eleven-meter waves.

Emission

Emission is the opposite of absorption, it is when an object emits radiation. Objects tend to emit amounts and wavelengths of radiation depending on their "black body" emission curves, therefore hotter objects tend to emit more radiation, with shorter wavelengths. Colder objects emit less radiation, with longer wavelengths. For example, the Sun is approximately 6,000 K (5,730 °C; 10,340 °F), its radiation peaks near 500 nm, and is visible to the human eye. The Earth is approximately 290 K (17 °C; 62 °F), so its radiation peaks near 10,000 nm, and is much too long to be visible to humans.

Because of its temperature, the atmosphere emits infrared radiation. For example, on clear nights the Earth's surface cools down faster than on cloudy nights. This is because clouds (H2O) are strong absorbers and emitters of infrared radiation. This is also why it becomes colder at night at higher elevations. The atmosphere acts as a "blanket" to limit the amount of radiation the Earth loses into space. The greenhouse effect is directly related to this absorption and emission (or "blanket") effect. Some chemicals in the atmosphere absorb and emit infrared radiation, but do not interact with sunlight in the visible spectrum. Common examples of these chemicals are CO2 and H2O. If there are too much of these greenhouse gases, sunlight heats the Earth's surface, but the gases block the infrared radiation from exiting back to space. This imbalance causes the Earth to warm, and thus climate change.

Refractive Index

The refractive index of air is close to, but just greater than 1. Systematic variations in refractive index can lead to the bending of light rays over long optical paths. One example is that, under some circumstances, observers onboard ships can see other vessels just over the horizon because light is refracted in the same direction as the curvature of the Earth's surface.

The refractive index of air depends on temperature, giving rise to refraction effects when the temperature gradient is large. An example of such effects is the mirage.

Circulation

Atmospheric circulation is the large-scale movement of air through the troposphere, and the means (with ocean circulation) by which heat is distributed around the Earth. The large-scale structure of the atmospheric circulation varies from year to year, but the basic structure remains fairly constant as it is determined by the Earth's rotation rate and the difference in solar radiation between the equator and poles.

Evolution of Earth's atmosphere

Earliest Atmosphere

The outgassings of the Earth were stripped away by solar winds early in the history of the planet until a steady state was established, the first atmosphere. Based on today's volcanic evidence, this atmosphere would have contained 60% hydrogen, 20% oxygen (mostly in the form of water vapor), 10% carbon dioxide, 5 to 7% hydrogen sulfide, and smaller amounts of nitrogen, carbon monoxide, free hydrogen, methane and inert gases. A major rainfall led to the buildup of a vast ocean, enriching the other agents, first carbon dioxide and later nitrogen and inert gases. A major part of carbon dioxide exhalations were soon dissolved in water and built up carbonate sediments.

Second Atmosphere

Water-related sediments have been found dating from as early as 3.8 billion years ago. About 3.4 billion years ago, nitrogen was the major part of the then stable "second atmosphere". An influence of life has to be taken into account rather soon in the history of the atmosphere, since hints of early life forms are to be found as early as 3.5 billion years ago. The fact that this is not perfectly in line with the 30% lower solar radiance (compared to today) of the early Sun has been described as the "faint young Sun paradox".

The geological record however shows a continually relatively warm surface during the complete early temperature record of the Earth with the exception of one cold glacial phase about 2.4 billion years ago. In the late Archaean eon an oxygen-containing atmosphere began to develop, apparently from photosynthesizing algae which have been found as stromatolite fossils from 2.7 billion years ago. The early basic carbon isotopy (isotope ratio proportions) is very much in line with what is found today, suggesting that the fundamental features of thecarbon cycle were established as early as 4 billion years ago.

Third Atmosphere

The accretion of continents about 3.5 billion years ago added plate tectonics, constantly rearranging the continents and also shaping long-term climate evolution by allowing the transfer of carbon dioxide to large land-based carbonate storages. Free oxygen did not exist until about 1.7 billion years ago and this can be seen with the development of the red beds and the end of the banded iron formations. This signifies a shift from a reducing atmosphere to an oxidising atmosphere. O2 showed major ups and downs until reaching a steady state of more than 15%. The following time span was the Phanerozoic eon, during which oxygen-breathing metazoan life forms began to appear.

Currently, anthropogenic greenhouse gases are increasing in the atmosphere. According to the Intergovernmental Panel on Climate Change, this increase is the main cause of global warming.

Atmospheric Stability

In order to have cloud formation, the air must be unstable. Stable air means air does not want to rise up, cool, and condense thus weather conditions tend to be clear skies. Unstable air means the air wants to rise up, cool, and condense into clouds and potential storms. So why does air rise to begin with?

Convection occurs when air rises much like a hot air balloon. Because of albedo, some areas on the ground can get heated up more than other areas. Where the land heats up more, the air above also warms, becomes less dense, and rises. If the air rises high enough, it may cool and condense to create clouds and possible thunderstorms.

Orographic uplift occurs when air must rise over a mountain range. As the air rises over the mountain, the moisture within it may begin to cool and condense to form thunderstorms. Often times with orographic uplift, one side of the mountain will be very moist from the storms, while the other side is very dry. The dry side of the mountain is called the rainshadow effect. Later we will discuss how this process can generate what is called dry thunderstorms and wildfires.

Picture Image taken by NASA's MODIS satellite.

Convergence occurs when air is forced to rise because of low pressure above causing the rising air to cool and condense into clouds. One of the best examples of this is over Florida. Because Florida is a peninsula (surrounded by water on three sides), the land heats up more and the surrounding water. This causes the air above the land to rise. To replace this rising air cool, moist air from the Gulf of Mexico and Atlantic Ocean converges inward over Florida. Now this moist air is heated by the land and is forced upward to create powerful thunderstorms. In fact, Florida has more thunderstorms and lightning than any other state in the nation. Another good example of convergence is the eye of a hurricane.

Finally weather fronts, such as cold fronts can force air to rise. A cold front occurs when a cold, dense air mass is replacing a warm, lighter air mass. The cold air mass plows through forcing the warmer air mass upward where is might cool and condense into clouds. But more on this in a minute.

Coriolis Effect

All free moving objects appear to be deflected to the right in the Northern Hemisphere (and to the left in the Southern Hemisphere) because of the rotation of the Earth. This apparent deflection is called the Coriolis effect. Think of it this way. If you wanted to through a football from Salt Lake City, Utah to Las Vegas, Nevada, you could not throw it from point A to point B. Why? Because as the football is moving toward Las Vegas, the Earth underneath is rotating. By the time the football gets to where Las Vegas was when threw the football, the city that never sleeps has moved off. Thus, if you want to throw a football from Salt Lake City to Las Vegas you would need to grow it ahead of the city in hopes that the football and Las Vegas meet up at the same time.

Besides a football, what else is considered a "free moving" object. Free moving objects include airplanes, ocean currents, clouds, wind, hurricanes, or anything else that is not attached to the earth's surface. So how does this influence the atmosphere? As winds flow from high to low pressure, it deflects to the right (in the Northern Hemisphere) because of the earth's rotation. This deflection is nonexistent at the equator and progressively gets stronger towards the poles. Because of this, hurricanes can only form 5-20 degrees north or south of the equator. Near the equator, the water is warm enough but the Coriolis effect is too weak to make the hurricane rotate. Beyond 20 degrees latitude, the Coriolis effect is strong enough, but the waters are too cold.

Picture

Air Masses and Weather Fronts

An air mass is a large body of air covering a relatively wide area and exhibiting horizontally uniform properties of moisture and temperature. The homogeneous of air masses comes from its source region. To create an air mass, the source region's surface must be fairly flat and uniform in composition (i.e. oceans, deserts, glaciers, etc.), but not a combination. If the air mass stays long enough within the source region, it will begin to develop the characteristics of that source region. An air mass is classified by their temperature and moisture and are identified by using a "letter code" system.

The first letter is always lower case and determines the moisture content within the air mass.

m (maritime) and is moist

c (continental) and is dry

The second letter is always a capital and determines latitude.

E (equatorial) and is very warm

T (tropical) and is warm

P (polar) and is cold

A (arctic) and is very cold

A weather front is a transition zone between two air masses of different densities, temperatures and/or moisture. When two air masses come into contact with each other, they do not like to mix well because of their different densities - much like water and oil. Along a weather front, the warmer, less dense air rises over the cooler, denser air to form clouds. There are several types of weather fronts, but we will only focus on two: cold fronts and warm fronts.

8

Ecology

Ecology is the study of the relationships between living organisms, including humans, and their physical environment. Ecology considers organisms at the individual, population, community, ecosystem, and biosphere level. Ecology overlaps with the closely related sciences of biogeography, evolutionary biology, genetics, ethology, and natural history. Ecology is a branch of biology, and it is not synonymous with environmentalism.

Among other things, ecology is the study of:

Life processes, antifragility, interactions, and adaptations

The movement of materials and energy through living communities

The successional development of ecosystems

Cooperation, competition, and predation within and between species

The abundance, biomass, and distribution of organisms in the context of the environment

Patterns of biodiversity and its effect on ecosystem processes

Ecology has practical applications in conservation biology, wetland management, natural resource management (agroecology, agriculture, forestry, agroforestry, fisheries), urban planning (urban ecology), community health, economics, basic and applied science, and human social interaction (human ecology).

The word ecology (German: Ökologie) was coined in 1866 by the German scientist Ernst Haeckel, and it became a rigorous science in the late 19th century. Evolutionary concepts relating to adaptation and natural selection are cornerstones of modern ecological theory.

Ecosystems are dynamically interacting systems of organisms, the communities they make up, and the non-living (abiotic) components of their environment. Ecosystem processes, such as primary production, nutrient cycling, and niche construction, regulate the flux of energy and matter through an environment. Ecosystems have biophysical feedback mechanisms that moderate processes acting on living (biotic) and abiotic components of the planet. Ecosystems sustain life-supporting functions and provide ecosystem services like biomass production (food, fuel, fiber, and medicine), the regulation of climate, global biogeochemical cycles, water filtration, soil formation, erosion control, flood protection, and many other natural features of scientific, historical, economic, or intrinsic value.

Levels, scope, and scale of organization

The scope of ecology contains a wide array of interacting levels of organization spanning micro-level (e.g., cells) to a planetary scale (e.g., biosphere) phenomena. Ecosystems, for example, contain abiotic resources and interacting life forms (i.e., individual organisms that aggregate into populations which aggregate into distinct ecological communities). Ecosystems are dynamic, they do not always follow a linear successional path, but they are always changing, sometimes rapidly and sometimes so slowly that it can take thousands of years for ecological processes to bring about certain successional stages of a forest. An ecosystem's area can vary greatly, from tiny to vast. A single tree is of little consequence to the classification of a forest ecosystem, but critically relevant to organisms living in and on it.

Several generations of an aphid population can exist over the lifespan of a single leaf. Each of those aphids, in turn, support diverse bacterial communities.

The nature of connections in ecological communities cannot be explained by knowing the details of each species in isolation, because the emergent pattern is neither revealed nor predicted until the ecosystem is studied as an integrated whole.

Some ecological principles, however, do exhibit collective properties where the sum of the components explain the properties of the whole, such as birth rates of a population being equal to the sum of individual births over a designated time frame.

The main subdisciplines of ecology, population (or community) ecology and ecosystem ecology, exhibit a difference not only of scale but also of two contrasting paradigms in the field. The former focuses on organisms' distribution and abundance, while the latter focuses on materials and energy fluxes.

Hierarchy

The scale of ecological dynamics can operate like a closed system, such as aphids migrating on a single tree, while at the same time remain open with regard to broader scale influences, such as atmosphere or climate. Hence, ecologists classify ecosystems hierarchically by analyzing data collected from finer scale units, such as vegetation associations, climate, and soil types, and integrate this information to identify emergent patterns of uniform organization and processes that operate on local to regional, landscape, and chronological scales.

To structure the study of ecology into a conceptually manageable framework, the biological world is organized into a nested hierarchy, ranging in scale from genes, to cells, to tissues, to organs, to organisms, to species, to populations, to communities, to ecosystems, to biomes, and up to the level of the biosphere.

This framework forms a panarchy and exhibits non-linear behaviors; this means that "effect and cause are disproportionate, so that small changes to critical variables, such as the number of nitrogen fixers, can lead to disproportionate, perhaps irreversible, changes in the system properties."

Biodiversity

Biodiversity (an abbreviation of "biological diversity") describes the diversity of life from genes to ecosystems and spans every level of biological organization. The term has several interpretations, and there are many ways to index, measure, characterize, and represent its complex organization.

Biodiversity includes species diversity, ecosystem diversity, and genetic diversity and scientists are interested in the way that this diversity affects the complex ecological processes operating at and among these respective levels.

Biodiversity plays an important role in ecosystem services which by definition maintain and improve human quality of life.

Conservation priorities and management techniques require different approaches and considerations to address the full ecological scope of biodiversity. Natural capital that supports populations is critical for maintaining ecosystem services and species migration (e.g., riverine fish runs and avian insect control) has been implicated as one mechanism by which those service losses are experienced.

An understanding of biodiversity has practical applications for species and ecosystem-level conservation planners as they make management recommendations to consulting firms, governments, and industry.

Habitat

The habitat of a species describes the environment over which a species is known to occur and the type of community that is formed as a result.

More specifically, "habitats can be defined as regions in environmental space that are composed of multiple dimensions, each representing a biotic or abiotic environmental variable; that is, any component or characteristic of the environment related directly (e.g. forage biomass and quality) or indirectly (e.g. elevation) to the use of a location by the animal."

For example, a habitat might be an aquatic or terrestrial environment that can be further categorized as a montane or alpine ecosystem. Habitat shifts provide important evidence of competition in nature where one population changes relative to the habitats that most other individuals of the species occupy. For example, one population of a species of tropical lizard (Tropidurus hispidus) has a flattened body relative to the main populations that live in open savanna. The population that lives in an isolated rock outcrop hides in crevasses where its flattened body offers a selective advantage. Habitat shifts also occur in the developmental life history of amphibians, and in insects that transition from aquatic to terrestrial habitats. Biotope and habitat are sometimes used interchangeably, but the former applies to a community's environment, whereas the latter applies to a species' environment.

Niche

The ecological niche is a central concept in the ecology of organisms and is sub-divided into the fundamental and the realized niche. The fundamental niche is the set of environmental conditions under which a species is able to persist. The realized niche is the set of environmental plus ecological conditions under which a species persists.

The Hutchinsonian niche is defined more technically as a "Euclidean hyperspace whose dimensions are defined as environmental variables and whose size is a function of the number of values that the environmental values may assume for which an organism has positive fitness.

Biogeographical patterns and range distributions are explained or predicted through knowledge of a species' traits and niche requirements.

Species have functional traits that are uniquely adapted to the ecological niche. A trait is a measurable property, phenotype, or characteristic of an organism that may influence its survival. Genes play an important role in the interplay of development and environmental expression of traits.

Resident species evolve traits that are fitted to the selection pressures of their local environment. This tends to afford them a competitive advantage and discourages similarly adapted species from having an overlapping geographic range. The competitive exclusion principle states that two species cannot coexist indefinitely by living off the same limiting resource; one will always out-compete the other. When similarly adapted species overlap geographically, closer inspection reveals subtle ecological differences in their habitat or dietary requirements.

Some models and empirical studies, however, suggest that disturbances can stabilize the co-evolution and shared niche occupancy of similar species inhabiting species-rich communities.

The habitat plus the niche is called the ecotope, which is defined as the full range of environmental and biological variables affecting an entire species.

Niche construction

Organisms are subject to environmental pressures, but they also modify their habitats. The regulatory feedback between organisms and their environment can affect conditions from local (e.g., a beaver pond) to global scales, over time and even after death, such as decaying logs or silica skeleton deposits from marine organisms.

The process and concept of ecosystem engineering are related to niche construction, but the former relates only to the physical modifications of the habitat whereas the latter also considers the evolutionary implications of physical changes to the environment and the feedback this causes on the process of natural selection. Ecosystem engineers are defined as: "organisms that directly or indirectly modulate the availability of resources to other species, by causing physical state changes in biotic or abiotic materials. In so doing they modify, maintain and create habitats."

The ecosystem engineering concept has stimulated a new appreciation for the influence that organisms have on the ecosystem and evolutionary process. The term "niche construction" is more often used in reference to the under-appreciated feedback mechanisms of natural selection imparting forces on the abiotic niche.

An example of natural selection through ecosystem engineering occurs in the nests of social insects, including ants, bees, wasps, and termites. There is an emergent homeostasis or homeorhesis in the structure of the nest that regulates, maintains and defends the physiology of the entire colony. Termite mounds, for example, maintain a constant internal temperature through the design of air-conditioning chimneys. The structure of the nests themselves is subject to the forces of natural selection. Moreover, a nest can survive over successive generations, so that progeny inherit both genetic material and a legacy niche that was constructed before their time.

Biome

Biomes are larger units of organization that categorize regions of the Earth's ecosystems, mainly according to the structure and composition of vegetation.

There are different methods to define the continental boundaries of biomes dominated by different functional types of vegetative communities that are limited in distribution by climate, precipitation, weather and other environmental variables. Biomes include tropical rainforest, temperate broadleaf and mixed forest, temperate deciduous forest, taiga, tundra, hot desert, and polar desert.

Other researchers have recently categorized other biomes, such as the human and oceanic microbiomes. To a microbe, the human body is a habitat and a landscape. Microbiomes were discovered largely through advances in molecular genetics, which have revealed a hidden richness of microbial diversity on the planet. The oceanic microbiome plays a significant role in the ecological biogeochemistry of the planet's oceans.

Biosphere

The largest scale of ecological organization is the biosphere: the total sum of ecosystems on the planet. Ecological relationships regulate the flux of energy, nutrients, and climate all the way up to the planetary scale. For example, the dynamic history of the planetary atmosphere's CO2 and O2 composition has been affected by the biogenic flux of gases coming from respiration and photosynthesis, with levels fluctuating over time in relation to the ecology and evolution of plants and animals. Ecological theory has also been used to explain self-emergent regulatory phenomena at the planetary scale: for example, the Gaia hypothesis is an example of holism applied in ecological theory. The Gaia hypothesis states that there is an emergent feedback loop generated by the metabolism of living organisms that maintains the core temperature of the Earth and atmospheric conditions within a narrow self-regulating range of tolerance.

Population ecology

Population ecology studies the dynamics of species populations and how these populations interact with the wider environment.

A population consists of individuals of the same species that live, interact, and migrate through the same niche and habitat.

A primary law of population ecology is the Malthusian growth model which states, "a population will grow (or decline) exponentially as long as the environment experienced by all individuals in the population remains constant.":?18? Simplified population models usually start with four variables: death, birth, immigration, and emigration.

Metapopulations and migration

The concept of metapopulations was defined in 1969 as "a population of populations which go extinct locally and recolonize".:?105? Metapopulation ecology is another statistical approach that is often used in conservation research. Metapopulation models simplify the landscape into patches of varying levels of quality, and metapopulations are linked by the migratory behaviours of organisms. Animal migration is set apart from other kinds of movement because it involves the seasonal departure and return of individuals from a habitat. Migration is also a population-level phenomenon, as with the migration routes followed by plants as they occupied northern post-glacial environments. Plant ecologists use pollen records that accumulate and stratify in wetlands to reconstruct the timing of plant migration and dispersal relative to historic and contemporary climates. These migration routes involved an expansion of the range as plant populations expanded from one area to another. There is a larger taxonomy of movement, such as commuting, foraging, territorial behaviour, stasis, and ranging. Dispersal is usually distinguished from migration because it involves the one-way permanent movement of individuals from their birth population into another population.

In metapopulation terminology, migrating individuals are classed as emigrants (when they leave a region) or immigrants (when they enter a region), and sites are classed either as sources or sinks. A site is a generic term that refers to places where ecologists sample populations, such as ponds or defined sampling areas in a forest. Source patches are productive sites that generate a seasonal supply of juveniles that migrate to other patch locations. Sink patches are unproductive sites that only receive migrants; the population at the site will disappear unless rescued by an adjacent source patch or environmental conditions become more favourable. Metapopulation models examine patch dynamics over time to answer potential questions about spatial and demographic ecology. The ecology of metapopulations is a dynamic process of extinction and colonization. Small patches of lower quality (i.e., sinks) are maintained or rescued by a seasonal influx of new immigrants. A dynamic metapopulation structure evolves from year to year, where some patches are sinks in dry years and are sources when conditions are more favourable. Ecologists use a mixture of computer models and field studies to explain metapopulation structure.

Community ecology

Community ecology is the study of the interactions among a collection of species that inhabit the same geographic area. Community ecologists study the determinants of patterns and processes for two or more interacting species. Research in community ecology might measure species diversity in grasslands in relation to soil fertility. It might also include the analysis of predator-prey dynamics, competition among similar plant species, or mutualistic interactions between crabs and corals.

Ecosystem ecology

Ecosystems may be habitats within biomes that form an integrated whole and a dynamically responsive system having both physical and biological complexes. Ecosystem ecology is the science of determining the fluxes of materials (e.g. carbon, phosphorus) between different pools (e.g., tree biomass, soil organic material). Ecosystem ecologists attempt to determine the underlying causes of these fluxes. Research in ecosystem ecology might measure primary production (g C/m^2) in a wetland in relation to decomposition and consumption rates (g C/m^2/y). This requires an understanding of the community connections between plants (i.e., primary producers) and the decomposers (e.g., fungi and bacteria),

The underlying concept of an ecosystem can be traced back to 1864 in the published work of George Perkins Marsh ("Man and Nature"). Within an ecosystem, organisms are linked to the physical and biological components of their environment to which they are adapted. Ecosystems are complex adaptive systems where the interaction of life processes form self-organizing patterns across different scales of time and space. Ecosystems are broadly categorized as terrestrial, freshwater, atmospheric, or marine. Differences stem from the nature of the unique physical environments that shapes the biodiversity within each. A more recent addition to ecosystem ecology are technoecosystems, which are affected by or primarily the result of human activity.

Food webs

A food web is the archetypal ecological network. Plants capture solar energy and use it to synthesize simple sugars during photosynthesis. As plants grow, they accumulate nutrients and are eaten by grazing herbivores, and the energy is transferred through a chain of organisms by consumption. The simplified linear feeding pathways that move from a basal trophic species to a top consumer is called the food chain. The larger interlocking pattern of food chains in an ecological community creates a complex food web. Food webs are a type of concept map or a heuristic device that is used to illustrate and study pathways of energy and material flows.

Food webs are often limited relative to the real world. Complete empirical measurements are generally restricted to a specific habitat, such as a cave or a pond, and principles gleaned from food web microcosm studies are extrapolated to larger systems. Feeding relations require extensive investigations into the gut contents of organisms, which can be difficult to decipher, or stable isotopes can be used to trace the flow of nutrient diets and energy through a food web. Despite these limitations, food webs remain a valuable tool in understanding community ecosystems.

Food webs exhibit principles of ecological emergence through the nature of trophic relationships: some species have many weak feeding links (e.g., omnivores) while some are more specialized with fewer stronger feeding links (e.g., primary predators). Theoretical and empirical studies identify non-random emergent patterns of few strong and many weak linkages that explain how ecological communities remain stable over time. Food webs are composed of subgroups where members in a community are linked by strong interactions, and the weak interactions occur between these subgroups. This increases food web stability. Step by step lines or relations are drawn until a web of life is illustrated.

Trophic levels

A trophic level is "a group of organisms acquiring a considerable majority of its energy from the lower adjacent level (according to ecological pyramids) nearer the abiotic source.":?383? Links in food webs primarily connect feeding relations or trophism among species. Biodiversity within ecosystems can be organized into trophic pyramids, in which the vertical dimension represents feeding relations that become further removed from the base of the food chain up toward top predators, and the horizontal dimension represents the abundance or biomass at each level. When the relative abundance or biomass of each species is sorted into its respective trophic level, they naturally sort into a 'pyramid of numbers'.

Species are broadly categorized as autotrophs (or primary producers), heterotrophs (or consumers), and Detritivores (or decomposers). Autotrophs are organisms that produce their own food (production is greater than respiration) by photosynthesis or chemosynthesis. Heterotrophs are organisms that must feed on others for nourishment and energy (respiration exceeds production).

Heterotrophs can be further sub-divided into different functional groups, including primary consumers (strict herbivores), secondary consumers (carnivorous predators that feed exclusively on herbivores), and tertiary consumers (predators that feed on a mix of herbivores and predators). Omnivores do not fit neatly into a functional category because they eat both plant and animal tissues. It has been suggested that omnivores have a greater functional influence as predators because compared to herbivores, they are relatively inefficient at grazing.

Trophic levels are part of the holistic or complex systems view of ecosystems. Each trophic level contains unrelated species that are grouped together because they share common ecological functions, giving a macroscopic view of the system. While the notion of trophic levels provides insight into energy flow and top-down control within food webs, it is troubled by the prevalence of omnivory in real ecosystems. This has led some ecologists to "reiterate that the notion that species clearly aggregate into discrete, homogeneous trophic levels is fiction. Nonetheless, recent studies have shown that real trophic levels do exist, but "above the herbivore trophic level, food webs are better characterized as a tangled web of omnivores.

Keystone species

A keystone species is a species that is connected to a disproportionately large number of other species in the food-web. Keystone species have lower levels of biomass in the trophic pyramid relative to the importance of their role. The many connections that a keystone species holds means that it maintains the organization and structure of entire communities. The loss of a keystone species results in a range of dramatic cascading effects that alters trophic dynamics, other food web connections, and can cause the extinction of other species.

Sea otters (Enhydra lutris) are commonly cited as an example of a keystone species because they limit the density of sea urchins that feed on kelp. If sea otters are removed from the system, the urchins graze until the kelp beds disappear, and this has a dramatic effect on community structure. Hunting of sea otters, for example, is thought to have led indirectly to the extinction of the Steller's sea cow (Hydrodamalis gigas). While the keystone species concept has been used extensively as a conservation tool, it has been criticized for being poorly defined from an operational stance. It is difficult to experimentally determine what species may hold a keystone role in each ecosystem. Furthermore, food web theory suggests that keystone species may not be common, so it is unclear how generally the keystone species model can be applied.

Complexity

Complexity is understood as a large computational effort needed to piece together numerous interacting parts exceeding the iterative memory capacity of the human mind. Global patterns of biological diversity are complex. This biocomplexity stems from the interplay among ecological processes that operate and influence patterns at different scales that grade into each other, such as transitional areas or ecotones spanning landscapes. Complexity stems from the interplay among levels of biological organization as energy, and matter is integrated into larger units that superimpose onto the smaller parts. "What were wholes on one level become parts on a higher one.": Small scale patterns do not necessarily explain large scale phenomena, otherwise captured in the expression (coined by Aristotle) 'the sum is greater than the parts'.[E]

"Complexity in ecology is of at least six distinct types: spatial, temporal, structural, process, behavioral, and geometric. From these principles, ecologists have identified emergent and self-organizing phenomena that operate at different environmental scales of influence, ranging from molecular to planetary, and these require different explanations at each integrative level. Ecological complexity relates to the dynamic resilience of ecosystems that transition to multiple shifting steady-states directed by random fluctuations of history.

Long-term ecological studies provide important track records to better understand the complexity and resilience of ecosystems over longer temporal and broader spatial scales. These studies are managed by the International Long Term Ecological Network (LTER). The longest experiment in existence is the Park Grass Experiment, which was initiated in 1856. Another example is the Hubbard Brook study, which has been in operation since 1960.

Holism

Holism remains a critical part of the theoretical foundation in contemporary ecological studies. Holism addresses the biological organization of life that self-organizes into layers of emergent whole systems that function according to non-reducible properties. This means that higher-order patterns of a whole functional system, such as an ecosystem, cannot be predicted or understood by a simple summation of the parts. "New properties emerge because the components interact, not because the basic nature of the components is changed."

Ecological studies are necessarily holistic as opposed to reductionistic.

Holism has three scientific meanings or uses that identify with ecology: 1) the mechanistic complexity of ecosystems, 2) the practical description of patterns in quantitative reductionist terms where correlations may be identified but nothing is understood about the causal relations without reference to the whole system, which leads to 3) a metaphysical hierarchy whereby the causal relations of larger systems are understood without reference to the smaller parts. Scientific holism differs from mysticism that has appropriated the same term. An example of metaphysical holism is identified in the trend of increased exterior thickness in shells of different species. The reason for a thickness increase can be understood through reference to principles of natural selection via predation without the need to reference or understand the biomolecular properties of the exterior shells.

Relation to evolution

Ecology and evolutionary biology are considered sister disciplines of the life sciences. Natural selection, life history, development, adaptation, populations, and inheritance are examples of concepts that thread equally into ecological and evolutionary theory. Morphological, behavioural, and genetic traits, for example, can be mapped onto evolutionary trees to study the historical development of a species in relation to their functions and roles in different ecological circumstances. In this framework, the analytical tools of ecologists and evolutionists overlap as they organize, classify, and investigate life through common systematic principles, such as phylogenetics or the Linnaean system of taxonomy. The two disciplines often appear together, such as in the title of the journal Trends in Ecology and Evolution. There is no sharp boundary separating ecology from evolution, and they differ more in their areas of applied focus. Both disciplines discover and explain emergent and unique properties and processes operating across different spatial or temporal scales of organization.

While the boundary between ecology and evolution is not always clear, ecologists study the abiotic and biotic factors that influence evolutionary processes, and evolution can be rapid, occurring on ecological timescales as short as one generation.

All organisms can exhibit behaviours. Even plants express complex behaviour, including memory and communication. Behavioural ecology is the study of an organism's behaviour in its environment and its ecological and evolutionary implications. Ethology is the study of observable movement or behaviour in animals. This could include investigations of motile sperm of plants, mobile phytoplankton, zooplankton swimming toward the female egg, the cultivation of fungi by weevils, the mating dance of a salamander, or social gatherings of amoeba.

Adaptation is the central unifying concept in behavioural ecology. Behaviours can be recorded as traits and inherited in much the same way that eye and hair colour can. Behaviours can evolve by means of natural selection as adaptive traits conferring functional utilities that increases reproductive fitness.

Predator-prey interactions are an introductory concept into food-web studies as well as behavioural ecology. Prey species can exhibit different kinds of behavioural adaptations to predators, such as avoid, flee, or defend. Many prey species are faced with multiple predators that differ in the degree of danger posed. To be adapted to their environment and face predatory threats, organisms must balance their energy budgets as they invest in different aspects of their life history, such as growth, feeding, mating, socializing, or modifying their habitat. Hypotheses posited in behavioural ecology are generally based on adaptive principles of conservation, optimization, or efficiency.

For example, "[t]he threat-sensitive predator avoidance hypothesis predicts that prey should assess the degree of threat posed by different predators and match their behaviour according to current levels of risk" or "[t]he optimal flight initiation distance occurs where expected postencounter fitness is maximized, which depends on the prey's initial fitness, benefits obtainable by not fleeing, energetic escape costs, and expected fitness loss due to predation risk."

Elaborate sexual displays and posturing are encountered in the behavioural ecology of animals. The birds-of-paradise, for example, sing and display elaborate ornaments during courtship. These displays serve a dual purpose of signalling healthy or well-adapted individuals and desirable genes. The displays are driven by sexual selection as an advertisement of quality of traits among suitors.

Cognitive ecology

Cognitive ecology integrates theory and observations from evolutionary ecology and neurobiology, primarily cognitive science, in order to understand the effect that animal interaction with their habitat has on their cognitive systems and how those systems restrict behavior within an ecological and evolutionary framework. "Until recently, however, cognitive scientists have not paid sufficient attention to the fundamental fact that cognitive traits evolved under particular natural settings. With consideration of the selection pressure on cognition, cognitive ecology can contribute intellectual coherence to the multidisciplinary study of cognition." As a study involving the 'coupling' or interactions between organism and environment, cognitive ecology is closely related to enactivism, a field based upon the view that "...we must see the organism and environment as bound together in reciprocal specification and selection...".

Social ecology

Social-ecological behaviours are notable in the social insects, slime moulds, social spiders, human society, and naked mole-rats where eusocialism has evolved. Social behaviours include reciprocally beneficial behaviours among kin and nest mates and evolve from kin and group selection. Kin selection explains altruism through genetic relationships, whereby an altruistic behaviour leading to death is rewarded by the survival of genetic copies distributed among surviving relatives. The social insects, including ants, bees, and wasps are most famously studied for this type of relationship because the male drones are clones that share the same genetic make-up as every other male in the colony. In contrast, group selectionists find examples of altruism among non-genetic relatives and explain this through selection acting on the group; whereby, it becomes selectively advantageous for groups if their members express altruistic behaviours to one another. Groups with predominantly altruistic members survive better than groups with predominantly selfish members.

Coevolution

Ecological interactions can be classified broadly into a host and an associate relationship. A host is any entity that harbours another that is called the associate. Relationships within a species that are mutually or reciprocally beneficial are called mutualisms. Examples of mutualism include fungus-growing ants employing agricultural symbiosis, bacteria living in the guts of insects and other organisms, the fig wasp and yucca moth pollination complex, lichens with fungi and photosynthetic algae, and corals with photosynthetic algae. If there is a physical connection between host and associate, the relationship is called symbiosis. Approximately 60% of all plants, for example, have a symbiotic relationship with arbuscular mycorrhizal fungi living in their roots forming an exchange network of carbohydrates for mineral nutrients.

Indirect mutualisms occur where the organisms live apart. For example, trees living in the equatorial regions of the planet supply oxygen into the atmosphere that sustains species living in distant polar regions of the planet. This relationship is called commensalism because many others receive the benefits of clean air at no cost or harm to trees supplying the oxygen.

If the associate benefits while the host suffers, the relationship is called parasitism. Although parasites impose a cost to their host (e.g., via damage to their reproductive organs or propagules, denying the services of a beneficial partner), their net effect on host fitness is not necessarily negative and, thus, becomes difficult to forecast. Co-evolution is also driven by competition among species or among members of the same species under the banner of reciprocal antagonism, such as grasses competing for growth space. The Red Queen Hypothesis, for example, posits that parasites track down and specialize on the locally common genetic defense systems of its host that drives the evolution of sexual reproduction to diversify the genetic constituency of populations responding to the antagonistic pressure.

Biogeography

Biogeography (an amalgamation of biology and geography) is the comparative study of the geographic distribution of organisms and the corresponding evolution of their traits in space and time. The Journal of Biogeography was established in 1974. Biogeography and ecology share many of their disciplinary roots. For example, the theory of island biogeography, published by the Robert MacArthur and Edward O. Wilson in 1967 is considered one of the fundamentals of ecological theory.

Biogeography has a long history in the natural sciences concerning the spatial distribution of plants and animals. Ecology and evolution provide the explanatory context for biogeographical studies. Biogeographical patterns result from ecological processes that influence range distributions, such as migration and dispersal. and from historical processes that split populations or species into different areas. The biogeographic processes that result in the natural splitting of species explain much of the modern distribution of the Earth's biota. The splitting of lineages in a species is called vicariance biogeography and it is a sub-discipline of biogeography. There are also practical applications in the field of biogeography concerning ecological systems and processes. For example, the range and distribution of biodiversity and invasive species responding to climate change is a serious concern and active area of research in the context of global warming.

Selection theory

A population ecology concept is r/K selection theory,[D] one of the first predictive models in ecology used to explain life-history evolution. The premise behind the r/K selection model is that natural selection pressures change according to population density. For example, when an island is first colonized, density of individuals is low. The initial increase in population size is not limited by competition, leaving an abundance of available resources for rapid population growth. These early phases of population growth experience density-independent forces of natural selection, which is called r-selection. As the population becomes more crowded, it approaches the island's carrying capacity, thus forcing individuals to compete more heavily for fewer available resources. Under crowded conditions, the population experiences density-dependent forces of natural selection, called K-selection.

In the r/K-selection model, the first variable r is the intrinsic rate of natural increase in population size and the second variable K is the carrying capacity of a population.

Different species evolve different life-history strategies spanning a continuum between these two selective forces. An r-selected species is one that has high birth rates, low levels of parental investment, and high rates of mortality before individuals reach maturity. Evolution favours high rates of fecundity in r-selected species. Many kinds of insects and invasive species exhibit r-selected characteristics. In contrast, a K-selected species has low rates of fecundity, high levels of parental investment in the young, and low rates of mortality as individuals mature. Humans and elephants are examples of species exhibiting K-selected characteristics, including longevity and efficiency in the conversion of more resources into fewer offspring.

Molecular ecology

The important relationship between ecology and genetic inheritance predates modern techniques for molecular analysis. Molecular ecological research became more feasible with the development of rapid and accessible genetic technologies, such as the polymerase chain reaction (PCR). The rise of molecular technologies and the influx of research questions into this new ecological field resulted in the publication Molecular Ecology in 1992. Molecular ecology uses various analytical techniques to study genes in an evolutionary and ecological context. In 1994, John Avise also played a leading role in this area of science with the publication of his book, Molecular Markers, Natural History and Evolution. Newer technologies opened a wave of genetic analysis into organisms once difficult to study from an ecological or evolutionary standpoint, such as bacteria, fungi, and nematodes. Molecular ecology engendered a new research paradigm for investigating ecological questions considered otherwise intractable. Molecular investigations revealed previously obscured details in the tiny intricacies of nature and improved resolution into probing questions about behavioural and biogeographical ecology. For example, molecular ecology revealed promiscuous sexual behaviour and multiple male partners in tree swallows previously thought to be socially monogamous. In a biogeographical context, the marriage between genetics, ecology, and evolution resulted in a new sub-discipline called phylogeography.

Human ecology

Ecology is as much a biological science as it is a human science.

Human ecology is an interdisciplinary investigation into the ecology of our species. "Human ecology may be defined: (1) from a bioecological standpoint as the study of man as the ecological dominant in plant and animal communities and systems; (2) from a bioecological standpoint as simply another animal affecting and being affected by his physical environment; and (3) as a human being, somehow different from animal life in general, interacting with physical and modified environments in a distinctive and creative way. A truly interdisciplinary human ecology will most likely address itself to all three." The term was formally introduced in 1921, but many sociologists, geographers, psychologists, and other disciplines were interested in human relations to natural systems centuries prior, especially in the late 19th century.

The ecological complexities human beings are facing through the technological transformation of the planetary biome has brought on the Anthropocene. The unique set of circumstances has generated the need for a new unifying science called coupled human and natural systems that builds upon, but moves beyond the field of human ecology. Ecosystems tie into human societies through the critical and all-encompassing life-supporting functions they sustain. In recognition of these functions and the incapability of traditional economic valuation methods to see the value in ecosystems, there has been a surge of interest in social-natural capital, which provides the means to put a value on the stock and use of information and materials stemming from ecosystem goods and services. Ecosystems produce, regulate, maintain, and supply services of critical necessity and beneficial to human health (cognitive and physiological), economies, and they even provide an information or reference function as a living library giving opportunities for science and cognitive development in children engaged in the complexity of the natural world. Ecosystems relate importantly to human ecology as they are the ultimate base foundation of global economics as every commodity, and the capacity for exchange ultimately stems from the ecosystems on Earth.

Restoration and management

Ecology is an employed science of restoration, repairing disturbed sites through human intervention, in natural resource management, and in environmental impact assessments. Edward O. Wilson predicted in 1992 that the 21st century "will be the era of restoration in ecology". Ecological science has boomed in the industrial investment of restoring ecosystems and their processes in abandoned sites after disturbance. Natural resource managers, in forestry, for example, employ ecologists to develop, adapt, and implement ecosystem based methods into the planning, operation, and restoration phases of land-use. Ecological science is used in the methods of sustainable harvesting, disease, and fire outbreak management, in fisheries stock management, for integrating land-use with protected areas and communities, and conservation in complex geo-political landscapes.

Relation to the environment

The environment of ecosystems includes both physical parameters and biotic attributes. It is dynamically interlinked and contains resources for organisms at any time throughout their life cycle.

Like ecology, the term environment has different conceptual meanings and overlaps with the concept of nature. Environment "includes the physical world, the social world of human relations and the built world of human creation. The physical environment is external to the level of biological organization under investigation, including abiotic factors such as temperature, radiation, light, chemistry, climate and geology. The biotic environment includes genes, cells, organisms, members of the same species (conspecifics) and other species that share a habitat.

The distinction between external and internal environments, however, is an abstraction parsing life and environment into units or facts that are inseparable in reality. There is an interpenetration of cause and effect between the environment and life. The laws of thermodynamics, for example, apply to ecology by means of its physical state. With an understanding of metabolic and thermodynamic principles, a complete accounting of energy and material flow can be traced through an ecosystem. In this way, the environmental and ecological relations are studied through reference to conceptually manageable and isolated material parts. After the effective environmental components are understood through reference to their causes; however, they conceptually link back together as an integrated whole, or holocoenotic system as it was once called. This is known as the dialectical approach to ecology. The dialectical approach examines the parts but integrates the organism and the environment into a dynamic whole (or umwelt). Change in one ecological or environmental factor can concurrently affect the dynamic state of an entire ecosystem.

Disturbance and resilience

Ecosystems are regularly confronted with natural environmental variations and disturbances over time and geographic space. A disturbance is any process that removes biomass from a community, such as a fire, flood, drought, or predation. Disturbances occur over vastly different ranges in terms of magnitudes as well as distances and time periods, and are both the cause and product of natural fluctuations in death rates, species assemblages, and biomass densities within an ecological community. These disturbances create places of renewal where new directions emerge from the patchwork of natural experimentation and opportunity. Ecological resilience is a cornerstone theory in ecosystem management. Biodiversity fuels the resilience of ecosystems acting as a kind of regenerative insurance.

Metabolism and the early atmosphere

The Earth was formed approximately 4.5 billion years ago. As it cooled and a crust and oceans formed, its atmosphere transformed from being dominated by hydrogen to one composed mostly of methane and ammonia. Over the next billion years, the metabolic activity of life transformed the atmosphere into a mixture of carbon dioxide, nitrogen, and water vapor. These gases changed the way that light from the sun hit the Earth's surface and greenhouse effects trapped heat. There were untapped sources of free energy within the mixture of reducing and oxidizing gasses that set the stage for primitive ecosystems to evolve and, in turn, the atmosphere also evolved.

Throughout history, the Earth's atmosphere and biogeochemical cycles have been in a dynamic equilibrium with planetary ecosystems. The history is characterized by periods of significant transformation followed by millions of years of stability. The evolution of the earliest organisms, likely anaerobic methanogen microbes, started the process by converting atmospheric hydrogen into methane (4H2 + CO2 ? CH4 + 2H2O). Anoxygenic photosynthesis reduced hydrogen concentrations and increased atmospheric methane, by converting hydrogen sulfide into water or other sulfur compounds (for example, 2H2S + CO2 + hv ? CH2O + H2O + 2S). Early forms of fermentation also increased levels of atmospheric methane. The transition to an oxygen-dominant atmosphere (the Great Oxidation) did not begin until approximately 2.4–2.3 billion years ago, but photosynthetic processes started 0.3 to 1 billion years prior.

Radiation: heat, temperature and light

The biology of life operates within a certain range of temperatures. Heat is a form of energy that regulates temperature. Heat affects growth rates, activity, behaviour, and primary production. Temperature is largely dependent on the incidence of solar radiation. The latitudinal and longitudinal spatial variation of temperature greatly affects climates and consequently the distribution of biodiversity and levels of primary production in different ecosystems or biomes across the planet. Heat and temperature relate importantly to metabolic activity. Poikilotherms, for example, have a body temperature that is largely regulated and dependent on the temperature of the external environment. In contrast, homeotherms regulate their internal body temperature by expending metabolic energy.

There is a relationship between light, primary production, and ecological energy budgets. Sunlight is the primary input of energy into the planet's ecosystems. Light is composed of electromagnetic energy of different wavelengths. Radiant energy from the sun generates heat, provides photons of light measured as active energy in the chemical reactions of life, and also acts as a catalyst for genetic mutation. Plants, algae, and some bacteria absorb light and assimilate the energy through photosynthesis. Organisms capable of assimilating energy by photosynthesis or through inorganic fixation of H2S are autotrophs. Autotrophs—responsible for primary production—assimilate light energy which becomes metabolically stored as potential energy in the form of biochemical enthalpic bonds.

Physical environments

Water

Diffusion of carbon dioxide and oxygen is approximately 10,000 times slower in water than in air. When soils are flooded, they quickly lose oxygen, becoming hypoxic (an environment with O2 concentration below 2 mg/liter) and eventually completely anoxic where anaerobic bacteria thrive among the roots. Water also influences the intensity and spectral composition of light as it reflects off the water surface and submerged particles. Aquatic plants exhibit a wide variety of morphological and physiological adaptations that allow them to survive, compete, and diversify in these environments. For example, their roots and stems contain large air spaces (aerenchyma) that regulate the efficient transportation of gases (for example, CO2 and O2) used in respiration and photosynthesis. Salt water plants (halophytes) have additional specialized adaptations, such as the development of special organs for shedding salt and osmoregulating their internal salt (NaCl) concentrations, to live in estuarine, brackish, or oceanic environments. Anaerobic soil microorganisms in aquatic environments use nitrate, manganese ions, ferric ions, sulfate, carbon dioxide, and some organic compounds; other microorganisms are facultative anaerobes and use oxygen during respiration when the soil becomes drier. The activity of soil microorganisms and the chemistry of the water reduces the oxidation-reduction potentials of the water. Carbon dioxide, for example, is reduced to methane (CH4) by methanogenic bacteria. The physiology of fish is also specially adapted to compensate for environmental salt levels through osmoregulation. Their gills form electrochemical gradients that mediate salt excretion in salt water and uptake in fresh water.

Gravity

The shape and energy of the land are significantly affected by gravitational forces. On a large scale, the distribution of gravitational forces on the earth is uneven and influences the shape and movement of tectonic plates as well as influencing geomorphic processes such as orogeny and erosion. These forces govern many of the geophysical properties and distributions of ecological biomes across the Earth. On the organismal scale, gravitational forces provide directional cues for plant and fungal growth (gravitropism), orientation cues for animal migrations, and influence the biomechanics and size of animals. Ecological traits, such as allocation of biomass in trees during growth are subject to mechanical failure as gravitational forces influence the position and structure of branches and leaves. The cardiovascular systems of animals are functionally adapted to overcome the pressure and gravitational forces that change according to the features of organisms (e.g., height, size, shape), their behaviour (e.g., diving, running, flying), and the habitat occupied (e.g., water, hot deserts, cold tundra).

Pressure

Climatic and osmotic pressure places physiological constraints on organisms, especially those that fly and respire at high altitudes, or dive to deep ocean depths. These constraints influence vertical limits of ecosystems in the biosphere, as organisms are physiologically sensitive and adapted to atmospheric and osmotic water pressure differences. For example, oxygen levels decrease with decreasing pressure and are a limiting factor for life at higher altitudes. Water transportation by plants is another important ecophysiological process affected by osmotic pressure gradients. Water pressure in the depths of oceans requires that organisms adapt to these conditions. For example, diving animals such as whales, dolphins, and seals are specially adapted to deal with changes in sound due to water pressure differences. Differences between hagfish species provide another example of adaptation to deep-sea pressure through specialized protein adaptations.

Wind and turbulence

Turbulent forces in air and water affect the environment and ecosystem distribution, form, and dynamics. On a planetary scale, ecosystems are affected by circulation patterns in the global trade winds. Wind power and the turbulent forces it creates can influence heat, nutrient, and biochemical profiles of ecosystems. For example, wind running over the surface of a lake creates turbulence, mixing the water column and influencing the environmental profile to create thermally layered zones, affecting how fish, algae, and other parts of the aquatic ecosystem are structured. Wind speed and turbulence also influence evapotranspiration rates and energy budgets in plants and animals. Wind speed, temperature and moisture content can vary as winds travel across different land features and elevations. For example, the westerlies come into contact with the coastal and interior mountains of western North America to produce a rain shadow on the leeward side of the mountain. The air expands and moisture condenses as the winds increase in elevation; this is called orographic lift and can cause precipitation. This environmental process produces spatial divisions in biodiversity, as species adapted to wetter conditions are range-restricted to the coastal mountain valleys and unable to migrate across the xeric ecosystems (e.g., of the Columbia Basin in western North America) to intermix with sister lineages that are segregated to the interior mountain systems.

Soils

Soil is the living top layer of mineral and organic dirt that covers the surface of the planet. It is the chief organizing centre of most ecosystem functions, and it is of critical importance in agricultural science and ecology. The decomposition of dead organic matter (for example, leaves on the forest floor), results in soils containing minerals and nutrients that feed into plant production. The whole of the planet's soil ecosystems is called the pedosphere where a large biomass of the Earth's biodiversity organizes into trophic levels. Invertebrates that feed and shred larger leaves, for example, create smaller bits for smaller organisms in the feeding chain. Collectively, these organisms are the detritivores that regulate soil formation. Tree roots, fungi, bacteria, worms, ants, beetles, centipedes, spiders, mammals, birds, reptiles, amphibians, and other less familiar creatures all work to create the trophic web of life in soil ecosystems. Soils form composite phenotypes where inorganic matter is enveloped into the physiology of a whole community. As organisms feed and migrate through soils they physically displace materials, an ecological process called bioturbation. This aerates soils and stimulates heterotrophic growth and production. Soil microorganisms are influenced by and are fed back into the trophic dynamics of the ecosystem. No single axis of causality can be discerned to segregate the biological from geomorphological systems in soils. Paleoecological studies of soils places the origin for bioturbation to a time before the Cambrian period. Other events, such as the evolution of trees and the colonization of land in the Devonian period played a significant role in the early development of ecological trophism in soils.

Biogeochemistry and climate

Ecologists study and measure nutrient budgets to understand how these materials are regulated, flow, and recycled through the environment. This research has led to an understanding that there is global feedback between ecosystems and the physical parameters of this planet, including minerals, soil, pH, ions, water, and atmospheric gases. Six major elements (hydrogen, carbon, nitrogen, oxygen, sulfur, and phosphorus; H, C, N, O, S, and P) form the constitution of all biological macromolecules and feed into the Earth's geochemical processes. From the smallest scale of biology, the combined effect of billions upon billions of ecological processes amplify and ultimately regulate the biogeochemical cycles of the Earth. Understanding the relations and cycles mediated between these elements and their ecological pathways has significant bearing toward understanding global biogeochemistry.

The ecology of global carbon budgets gives one example of the linkage between biodiversity and biogeochemistry. It is estimated that the Earth's oceans hold 40,000 gigatonnes (Gt) of carbon, that vegetation and soil hold 2070 Gt, and that fossil fuel emissions are 6.3 Gt carbon per year. There have been major restructurings in these global carbon budgets during the Earth's history, regulated to a large extent by the ecology of the land. For example, through the early-mid Eocene volcanic outgassing, the oxidation of methane stored in wetlands, and seafloor gases increased atmospheric CO_2 (carbon dioxide) concentrations to levels as high as 3500 ppm.

In the Oligocene, from twenty-five to thirty-two million years ago, there was another significant restructuring of the global carbon cycle as grasses evolved a new mechanism of photosynthesis, C4 photosynthesis, and expanded their ranges. This new pathway evolved in response to the drop in atmospheric CO_2 concentrations below 550 ppm. The relative abundance and distribution of biodiversity alters the dynamics between organisms and their environment such that ecosystems can be both cause and effect in relation to climate change. Human-driven modifications to the planet's ecosystems (e.g., disturbance, biodiversity loss, agriculture) contributes to rising atmospheric greenhouse gas levels. Transformation of the global carbon cycle in the next century is projected to raise planetary temperatures, lead to more extreme fluctuations in weather, alter species distributions, and increase extinction rates. The effect of global warming is already being registered in melting glaciers, melting mountain ice caps, and rising sea levels. Consequently, species distributions are changing along waterfronts and in continental areas where migration patterns and breeding grounds are tracking the prevailing shifts in climate. Large sections of permafrost are also melting to create a new mosaic of flooded areas having increased rates of soil decomposition activity that raises methane (CH_4) emissions. There is concern over increases in atmospheric methane in the context of the global carbon cycle, because methane is a greenhouse gas that is 23 times more effective at absorbing long-wave radiation than CO_2 on a 100-year time scale. Hence, there is a relationship between global warming, decomposition and respiration in soils and wetlands producing significant climate feedbacks and globally altered biogeochemical cycles.

9

Physical Features of the Desert

A desert is a barren area of landscape where little precipitation occurs and, consequently, living conditions are hostile for plant and animal life. The lack of vegetation exposes the unprotected surface of the ground to the processes of denudation. About one-third of the land surface of the Earth is arid or semi-arid. This includes much of the polar regions, where little precipitation occurs, and which are sometimes called polar deserts or "cold deserts". Deserts can be classified by the amount of precipitation that falls, by the temperature that prevails, by the causes of desertification or by their geographical location.

Deserts are formed by weathering processes as large variations in temperature between day and night put strains on the rocks, which consequently break in pieces. Although rain seldom occurs in deserts, there are occasional downpours that can result in flash floods. Rain falling on hot rocks can cause them to shatter, and the resulting fragments and rubble strewn over the desert floor are further eroded by the wind. This picks up particles of sand and dust, which can remain airborne for extended periods – sometimes causing the formation of sand storms or dust storms. Wind-blown sand grains striking any solid object in their path can abrade the surface. Rocks are smoothed down, and the wind sorts sand into uniform deposits. The grains end up as level sheets of sand or are piled high in billowing sand dunes. Other deserts are flat, stony plains where all the fine material has been blown away and the surface consists of a mosaic of smooth stones. These areas are known as desert pavements, and little further erosion takes place. Other desert features include rock outcrops, exposed bedrock and clays once deposited by flowing water. Temporary lakes may form and salt pans may be left when waters evaporate. There may be underground sources of water, in the form of springs and seepages from aquifers. Where these are found, oases can occur.

Plants and animals living in the desert need special adaptations to survive in the harsh environment. Plants tend to be tough and wiry with small or no leaves, water-resistant cuticles, and often spines to deter herbivory. Some annual plants germinate, bloom and die in the course of a few weeks after rainfall, while other long-lived plants survive for years and have deep root systems able to tap underground moisture. Animals need to keep cool and find enough food and water to survive. Many are nocturnal, and stay in the shade or underground during the heat of the day. They tend to be efficient at conserving water, extracting most of their needs from their food and concentrating their urine. Some animals remain in a state of dormancy for long periods, ready to become active again during the rare rainfall. They then reproduce rapidly while conditions are favorable before returning to dormancy.

People have struggled to live in deserts and the surrounding semi-arid lands for millennia. Nomads have moved their flocks and herds to wherever grazing is available, and oases have provided opportunities for a more settled way of life. The cultivation of semi-arid regions encourages erosion of soil and is one of the causes of increased desertification. Desert farming is possible with the aid of irrigation, and the Imperial Valley in California provides an example of how previously barren land can be made productive by the import of water from an outside source. Many trade routes have been forged across deserts, especially across the Sahara, and traditionally were used by caravans of camels carrying salt, gold, ivory and other goods. Large numbers of slaves were also taken northwards across the Sahara. Some mineral extraction also takes place in deserts, and the uninterrupted sunlight gives potential for the capture of large quantities of solar energy.

Physical geography

A desert is a region of land that is very dry because it receives low amounts of precipitation (usually in the form of rain, but it may be snow, mist or fog), often has little coverage by plants, and in which streams dry up unless they are supplied by water from outside the area.

Deserts generally receive less than 250 mm (10 in) of precipitation each year.

The potential evapotranspiration may be large but (in the absence of available water) the actual evapotranspiration may be close to zero.

Semi-deserts are regions which receive between 250 and 500 mm (10 and 20 in) and when clad in grass, these are known as steppes.

Classification

Deserts are sometimes classified as "hot" and "cold" deserts. Cold deserts can be covered in snow or ice; frozen water unavailable to plant life. These are more commonly referred to as tundra if a short season of above-freezing temperatures is experienced, or as an ice cap if the temperature remains below freezing year-round, rendering the land almost completely lifeless.

In 1961, Peveril Meigs divided desert regions on Earth into three categories according to the amount of precipitation they received. In this now widely accepted system, extremely arid lands have at least 12 consecutive months without rainfall, arid lands have less than 250 mm (10 in) of annual rainfall, and semiarid lands have a mean annual precipitation of between 250 and 500 mm (10–20 in). Arid and extremely arid lands are deserts, and semiarid areas are generally referred to as steppes.

In some parts of the world, deserts are created by a rain shadow effect in which air masses lose much of their moisture as they move over a mountain range; other areas are arid by virtue of being very far from the nearest available sources of moisture.

Deserts are also classified by their geographical location and dominant weather pattern as trade wind, mid-latitude, rain shadow, coastal, monsoon, or polar deserts. Former desert areas presently in non-arid environments are paleodeserts.

Montane deserts are arid places with a very high altitude; the most prominent example is found north of the Himalayas, especially in Ladakh region of Jammu and Kashmir, in parts of the Kunlun Mountains and the Tibetan Plateau. Many locations within this category have elevations exceeding 3,000 meters (10,000 ft) and the thermal regime can be hemiboreal. These places owe their profound aridity (the average annual precipitation is often less than 40 mm or 1.5 in) to being very far from the nearest available sources of moisture. Montane deserts are normally cold.

Rain shadow deserts form when tall mountain ranges block clouds from reaching areas in the direction the wind is going. As the air moves over the mountains, it cools and moisture condenses, causing precipitation on the windward side. When that air reaches the leeward side, it is dry because it has lost the majority of its moisture, resulting in a desert. The air then warms, expands, and blows across the desert. The warm, desiccated air takes with it any remaining moisture in the desert.

Deserts take up about one third (33%) of the Earth's land surface. Hot deserts usually have a large diurnal and seasonal temperature range, with high daytime temperatures, and low nighttime temperatures (due to extremely low humidity). In hot deserts the temperature in the daytime can reach 45 °C/113 °F or higher in the summer, and dip to 0 °C/32 °F or lower at nighttime in the winter.

Water vapor in the atmosphere acts to trap long wave infrared radiation from the ground, and dry desert air is incapable of blocking sunlight during the day (due to absence of clouds) or trapping heat during the night. Thus, during daylight most of the sun's heat reaches the ground, and as soon as the sun sets the desert cools quickly by radiating its heat into space. Urban areas in deserts lack large (more than 14 °C/ 25 °F) daily temperature variations, partially due to the urban heat island effect.

Many deserts are formed by rain shadows; mountains blocking the path of precipitation to the desert (on the lee side of the mountain). Deserts are often composed of sand and rocky surfaces. Sand dunes called ergs and stony surfaces called hamada surfaces compose a minority of desert surfaces. Exposures of rocky terrain are typical, and reflect minimal soil development and sparseness of vegetation. The soil is rocky because of the low chemical weathering, and the relative absence of a humus fraction.

Bottomlands may be salt-covered flats. Eolian processes are major factors in shaping desert landscapes. Polar deserts (also seen as "cold deserts") have similar features, except the main form of precipitation is snow rather than rain. Antarctica is the world's largest cold desert (composed of about 98% thick continental ice sheet and 2% barren rock). Some of the barren rock is to be found in the so-called Dry Valleysof Antarctica that almost never get snow, which can have ice-encrusted saline lakes that suggest evaporation far greater than the rare snowfall due to the strong katabatic winds that evaporate even ice.

The largest hot desert is the Sahara in northern Africa, covering 9 million square kilometres and 12 countries.

Deserts sometimes contain valuable mineral deposits that were formed in the arid environment or that were exposed by erosion. Due to extreme and consistent dryness, some deserts are ideal places for natural preservation of artifacts and fossils.

Desert Features

Sand covers only about 20% of Earth's deserts. Most of the sand is in sand sheets and sand seas—vast regions of undulating dunes resembling ocean waves "frozen" in an instant of time. In general, there are five forms of deserts:

- Mountain and basin deserts
- Hamada deserts, which consist of plateau landforms
- Regs, which consist of rock pavements
- Ergs, which are formed by sand seas
- Intermontane Basins.

Nearly all desert surfaces are plains where eolian deflation—removal of fine-grained material by the wind—has exposed loose gravels consisting predominantly of pebbles but with occasional cobbles.

The remaining surfaces of arid lands are composed of exposed bedrock outcrops, desert soils, and fluvial deposits including alluvial fans,playas, desert lakes, and oases. Bedrock outcrops occur as small mountains surrounded by extensive erosional plains.

Several different types of dunes exist. Barchan dunes are produced by strong winds blowing across a level surface and are crescent-shaped. Longitudinal or seif dunes are dunes that are parallel to a strong wind that blows in one general direction. Transverse dunes run at a right angle to the constant wind direction. Star dunes are star-shaped and have several ridges that spread out around a point.

Oases are vegetated areas moistened by springs, wells, or by irrigation. Many are artificial. Oases are often the only places in deserts that support crops and permanent habitation.

Ecology and biogeography

Deserts and semi-deserts are home to ecosystems with low or very low biomass and primary productivity in arid or semi-arid climates. They are mostly found in subtropical high-pressure belts and major continental rain shadows. Primary productivity depends on low densities of small photoautotrophs that sustain a sparse trophic network. Plant growth is limited by rainfall, temperature extremes and desiccating winds. Deserts have strong temporal variability in the availability of resources due to the total amount of annual rainfall and the size of individual rainfall events. Resources are often ephemeral or episodic, and this triggers sporadic animal movements and 'pulse and reserve' or 'boom-bust' ecosystem dynamics. Erosion and sedimentation are high due to the sparse vegetation cover and the activities of large mammals and people. Plants and animals in deserts are mostly adapted to extreme and prolonged water deficits, but their reproductive phenology often responds to short episodes of surplus. Competitive interactions are weak.

Flora

Plants face severe challenges in arid environments. Problems they need to solve include how to obtain enough water, how to avoid being eaten and how to reproduce.

Photosynthesis is the key to plant growth. It can only take place during the day as energy from the sun is required, but during the day, many deserts become very hot. Opening stomata to allow in the carbon dioxide necessary for the process causes evapotranspiration, and conservation of water is a top priority for desert vegetation. Some plants have resolved this problem by adopting crassulacean acid metabolism, allowing them to open their stomata during the night to allow CO2 to enter, and close them during the day, or by using C4 carbon fixation.

Many desert plants have reduced the size of their leaves or abandoned them altogether. Cacti are desert specialists, and in most species, the leaves have been dispensed with and the chlorophyll displaced into the trunks, the cellular structure of which has been modified to allow them to store water. When rain falls, the water is rapidly absorbed by the shallow roots and retained to allow them to survive until the next downpour, which may be months or years away. The giant saguaro cacti of the Sonoran Desert form "forests", providing shade for other plants and nesting places for desert birds. Saguaro grows slowly but may live for up to two hundred years. The surface of the trunk is folded like a concertina, allowing it to expand, and a large specimen can hold eight tons of water after a good downpour.

Cacti are present in both North and South America with a post-Gondwana origin. Other xerophytic plants have developed similar strategies by a process known as convergent evolution. They limit water loss by reducing the size and number of stomata, by having waxy coatings and hairy or tiny leaves. Some are deciduous, shedding their leaves in the driest season, and others curl their leaves up to reduce transpiration. Others store water in succulent leaves or stems or in fleshy tubers. Desert plants maximize water uptake by having shallow roots that spread widely, or by developing long taproots that reach down to deep rock strata for ground water. The saltbush in Australia has succulent leaves and secretes salt crystals, enabling it to live in saline areas. In common with cacti, many have developed spines to ward off browsing animals.

Some desert plants produce seed which lies dormant in the soil until sparked into growth by rainfall. With annuals, such plants grow with great rapidity and may flower and set seed within weeks, aiming to complete their development before the last vestige of water dries up. For perennial plants, reproduction is more likely to be successful if the seed germinates in a shaded position, but not so close to the parent plant as to be in competition with it. Some seed will not germinate until it has been blown about on the desert floor to scarify the seed coat. The seed of the mesquite tree, which grows in deserts in the Americas, is hard and fails to sprout even when planted carefully. When it has passed through the gut of a pronghorn it germinates readily, and the little pile of moist dung provides an excellent start to life well away from the parent tree. The stems and leaves of some plants lower the surface velocity of sand-carrying winds and protect the ground from erosion. Even small fungi and microscopic plant organisms found on the soil surface (so-called cryptobiotic soil) can be a vital link in preventing erosion and providing support for other living organisms. Cold deserts often have high concentrations of salt in the soil. Grasses and low shrubs are the dominant vegetation here and the ground may be covered with lichens. Most shrubs have spiny leaves and shed them in the coldest part of the year.

Fauna

Animals adapted to live in deserts are called xerocoles. There is no evidence that body temperature of mammals and birds is adaptive to the different climates, either of great heat or cold. In fact, with a very few exceptions, their basal metabolic rate is determined by body size, irrespective of the climate in which they live. Many desert animals (and plants) show especially clear evolutionary adaptations for water conservation or heat tolerance and so are often studied in comparative physiology, ecophysiology, and evolutionary physiology. One well-studied example is the specializations of mammalian kidneys shown by desert-inhabiting species. Many examples of convergent evolution have been identified in desert organisms, including between cacti and Euphorbia, kangaroo rats and jerboas, Phrynosoma and Moloch lizards.

Deserts present a very challenging environment for animals. Not only do they require food and water but they also need to keep their body temperature at a tolerable level. In many ways, birds are the ablest to do this of the higher animals. They can move to areas of greater food availability as the desert blooms after local rainfall and can fly to faraway waterholes. In hot deserts, gliding birds can remove themselves from the over-heated desert floor by using thermals to soar in the cooler air at great heights. In order to conserve energy, other desert birds run rather than fly. The cream-colored courser flits gracefully across the ground on its long legs, stopping periodically to snatch up insects. Like other desert birds, it is well-camouflaged by its coloring and can merge into the landscape when stationary. The sandgrouse is an expert at this and nests on the open desert floor dozens of kilometers (miles) away from the waterhole it needs to visit daily. Some small diurnal birds are found in very restricted localities where their plumage matches the color of the underlying surface. The desert lark takes frequent dust baths which ensures that it matches its environment.

Water and carbon dioxide are metabolic end products of oxidation of fats, proteins, and carbohydrates. Oxidising a gram of carbohydrate produces 0.60 grams of water; a gram of protein produces 0.41 grams of water; and a gram of fat produces 1.07 grams of water, making it possible for xerocoles to live with little or no access to drinking water. The kangaroo rat for example makes use of this water of metabolism and conserves water both by having a low basal metabolic rate and by remaining underground during the heat of the day, reducing loss of water through its skin and respiratory system when at rest. Herbivorous mammals obtain moisture from the plants they eat. Species such as the addax antelope, dik-dik, Grant's gazelle and oryx are so efficient at doing this that they apparently never need to drink. The camel is a superb example of a mammal adapted to desert life. It minimizes its water loss by producing concentrated urine and dry dung, and is able to lose 40% of its body weight through water loss without dying of dehydration. Carnivores can obtain much of their water needs from the body fluids of their prey. Many other hot desert animals are nocturnal, seeking out shade during the day or dwelling underground in burrows. At depths of more than 50 cm (20 in), these remain at between 30 to 32 °C (86 to 90 °F) regardless of the external temperature. Jerboas, desert rats, kangaroo rats and other small rodents emerge from their burrows at night and so do the foxes, coyotes, jackals and snakes that prey on them. Kangaroos keep cool by increasing their respiration rate, panting, sweating and moistening the skin of their forelegs with saliva. Mammals living in cold deserts have developed greater insulation through warmer body fur and insulating layers of fat beneath the skin. The arctic weasel has a metabolic rate that is two or three times as high as would be expected for an animal of its size. Birds have avoided the problem of losing heat through their feet by not attempting to maintain them at the same temperature as the rest of their bodies, a form of adaptive insulation. The emperor penguin has dense plumage, a downy under layer, an air insulation layer next to the skin and various thermoregulatory strategies to maintain its body temperature in one of the harshest environments on Earth.

Being ectotherms, reptiles are unable to live in cold deserts but are well-suited to hot ones. In the heat of the day in the Sahara, the temperature can rise to 50 °C (122 °F). Reptiles cannot survive at this temperature and lizards will be prostrated by heat at 45 °C (113 °F). They have few adaptations to desert life and are unable to cool themselves by sweating so they shelter during the heat of the day. In the first part of the night, as the ground radiates the heat absorbed during the day, they emerge and search for prey. Lizards and snakes are the most numerous in arid regions and certain snakes have developed a novel method of locomotion that enables them to move sidewards and navigate high sand-dunes. These include the horned viper of Africa and the sidewinder of North America, evolutionarily distinct but with similar behavioural patterns because of convergent evolution. Many desert reptiles are ambush predators and often bury themselves in the sand, waiting for prey to come within range.

Amphibians might seem unlikely desert-dwellers, because of their need to keep their skins moist and their dependence on water for reproductive purposes. In fact, the few species that are found in this habitat have made some remarkable adaptations. Most of them are fossorial, spending the hot dry months aestivating in deep burrows. While there they shed their skins a number of times and retain the remnants around them as a waterproof cocoon to retain moisture. In the Sonoran Desert, Couch's spadefoot toad spends most of the year dormant in its burrow. Heavy rain is the trigger for emergence and the first male to find a suitable pool calls to attract others. Eggs are laid and the tadpoles grow rapidly as they must reach metamorphosis before the water evaporates. As the desert dries out, the adult toads rebury themselves. The juveniles stay on the surface for a while, feeding and growing, but soon dig themselves burrows. Few make it to adulthood. The water holding frog in Australia has a similar life cycle and may aestivate for as long as five years if no rain falls. The Desert rain frog of Namibia is nocturnal and survives because of the damp sea fogs that roll in from the Atlantic.

Invertebrates, particularly arthropods, have successfully made their homes in the desert. Flies, beetles, ants, termites, locusts, millipedes, scorpions and spiders have hard cuticles which are impervious to water and many of them lay their eggs underground and their young develop away from the temperature extremes at the surface. The Saharan silver ant (Cataglyphis bombycina) uses a heat shock protein in a novel way and forages in the open during brief forays in the heat of the day. The long-legged darkling beetle in Namibia stands on its front legs and raises its carapace to catch the morning mist as condensate, funnelling the water into its mouth. Some arthropods make use of the ephemeral pools that form after rain and complete their life cycle in a matter of days. The desert shrimp does this, appearing "miraculously" in new-formed puddles as the dormant eggs hatch. Others, such as brine shrimps, fairy shrimps and tadpole shrimps, are cryptobiotic and can lose up to 92% of their bodyweight, rehydrating as soon as it rains and their temporary pools reappear.

Water

Atacama is the driest place on Earth and is virtually sterile because it is blocked from moisture on both sides by the Andes mountains and by the Chilean Coast Range. The cold Humboldt Current and the anticyclone of the Pacific are essential to keep the dry climate of the Atacama. The average rainfall in the Chilean region of Antofagasta is just 1 mm per year. Some weather stations in the Atacama have never received rain. Evidence suggests that the Atacama may not have had any significant rainfall from 1570 to 1971. It is so arid that mountains that reach as high as 6,885 meters (22,590 feet) are completely free of glaciers and, in the southern part from 25°S to 27°S, may have been glacier-free throughout the Quaternary, though permafrost extends down to an altitude of 4,400 meters and is continuous above 5,600 meters.

Rain does fall occasionally in deserts, and desert storms are often violent. A record 44 millimeters (1.7 in) of rain once fell within 3 hours in the Sahara. Large Saharan storms may deliver up to 1 millimeter per minute. Normally dry stream channels, called arroyos or wadis, can quickly fill after heavy rains, and flash floods make these channels dangerous.

Though little rain falls in deserts, deserts receive runoff from ephemeral, or short-lived, streams fed considerable quantities of sediment for a day or two. Although most deserts are in basins with closed or interior drainage, a few deserts are crossed by 'exotic' rivers that derive their water from outside the desert. Such rivers infiltrate soils and evaporate large amounts of water on their journeys through the deserts, but their volumes are such that they maintain their continuity.

The Nile River, the Colorado River, and the Yellow River are exotic rivers that flow through deserts to deliver their sediments to the sea. Deserts may also have underground springs, rivers, or reservoirs that lie close to the surface, or deep underground. Plants that have not completely adapted to sporadic rainfalls in a desert environment may tap into underground water sources that do not exceed the reach of their root systems.

While deserts are well-known for their lack of water, some groups have adapted ways to find water in this harsh environment. The Bedouin, for example, turn over half-buried stones just before dawn so dew forms on them.

Lakes form where rainfall or meltwater in interior drainage basins is sufficient. Desert lakes are generally shallow, temporary, and salty. Because these lakes are shallow and have a low bottom gradient, wind stress may cause the lake waters to move over many square kilometers. When small lakes dry up, they leave a salt crust or hardpan. The flat area of clay, silt, or sand encrusted with salt that forms is known as a playa or a sink. There are more than a hundred playas in North American deserts.

Most are relics of large lakes that existed during the last ice age about 12,000 years ago. Lake Bonneville was a 52,000-square-kilometer (20,000 mi^2) lake almost 300 meters (1000 ft) deep in Utah, Nevada, and Idaho during the Ice Age. Today the remnants of Lake Bonneville include Utah's Great Salt Lake, Utah Lake, and Sevier Lake. Because playas are arid landforms from a wetter past, they contain useful clues to climatic change.

When the occasional precipitation does occur, it erodes the desert rocks quickly. The flat terrains of hardpans and playas make them excellent racetracks and natural runways for airplanes and spacecraft. Ground-vehicle speed records have been established on the flat lakebeds of the Black Rock Desert in Nevada and Bonneville Speedway in Utah. Space shuttles and flight-test aircraft land on Rogers Lake Playa at Edwards Air Force Base in California.

Formation of Hot Deserts

There are four main, interlinked causes of hot deserts:

- The formation of the subtropical high-pressure cell.
- The rain shadow effect in the belt of easterly trade winds.
- The effect of the cold currents off the west coast of the continents at these latitudes.
- The depositing sands of a desert along its border into the fertile land

Hot deserts (like cold deserts) may result in average temperature cooling because they reflect more of the incoming light (their albedo is higher than that of water or forests).

Mineral Resources

Deserts may contain great amounts of mineral resources over their entire surface. This occurrence in minerals also determines the color. For example, the red color of many sand deserts is a result of the occurrence of laterite. Some mineral deposits are formed, improved, or preserved by geologic processes that occur in arid lands as a consequence of climate.Ground water leaches ore minerals and redeposits them in zones near the water table. This leaching process concentrates these minerals as ore that can be mined.

Evaporation in arid lands enriches mineral accumulation in their lakes. Lake beds known as playas may be sources of mineral deposits formed by evaporation. Water evaporating in closed basins precipitates minerals such as gypsum, salts (including sodium nitrate and sodium chloride), and borates. The minerals formed in these evaporite deposits depend on the composition and temperature of the saline waters at the time of deposition.

Significant evaporite resources occur in the Great Basin Desert of the United States, mineral deposits made famous by the "20-mule teams" that once hauled borax-laden wagons from Death Valley to the railroad. Boron, from borax and borate evaporites, is an essential ingredient in the manufacture of glass, enamel, agricultural chemicals, water softeners, and pharmaceuticals. Borates are mined from evaporite deposits at Searles Lake, California, and other desert locations. The total value of chemicals that have been produced from Searles Lake substantially exceeds US$1 billion.

The Atacama Desert of Chile is unique among the deserts of the world in its great abundance of saline minerals. Sodium nitrate has been mined for explosives and fertilizer in the Atacama since the middle of the 19th century. Nearly 3 million metric tons were mined during World War I.

Valuable minerals located in arid lands include copper in the United States, Chile, Peru, and Iran; iron and lead-zinc ore in Australia; and gold, silver, and uranium deposits in Australia and the United States. Nonmetallic mineral resources and rocks such as beryllium, mica, lithium, clays, pumice, and scoria also occur in arid regions. Sodium carbonate, sulfate, borate,nitrate, lithium, bromine, iodine, calcium, and strontium compounds come from sediments and near-surface brines formed by evaporation of inland bodies of water, often during geologically recent times.

The Green River Formation of Colorado, Wyoming, and Utah contains alluvial fan deposits and playa evaporites created in a huge lake whose level fluctuated for millions of years. Economically significant deposits of trona, a major source of sodium compounds, and thick layers of oil shale were created in the arid environment.

Some of the more productive petroleum areas on Earth are found in arid and semiarid regions of Africa and the Mideast, although the oil fields were originally formed in shallow marine environments. Recent climate change has placed these reservoirs in an arid environment. It's noteworthy that Ghawar, the world's largest and most productive oilfield is mostly under the Empty Quarter and Al-Dahna deserts.

Other oil reservoirs, however, are presumed to be eolian in origin and are presently found in humid environments. The Rotliegendes, a hydrocarbon reservoir in the North Sea, is associated with extensive evaporite deposits. Many of the major U.S. hydrocarbon resources may come from eolian sands. Ancient alluvial fan sequences may also be hydrocarbon reservoirs.

Weathering processes

Deserts usually have a large diurnal and seasonal temperature range, with high daytime temperatures falling sharply at night. The diurnal range may be as much as 20 to 30 °C (36 to 54 °F) and the rock surface experiences even greater temperature differentials.

During the day the sky is usually clear and most of the sun's radiation reaches the ground, but as soon as the sun sets, the desert cools quickly by radiating heat into space. In hot deserts, the temperature during daytime can exceed 45 °C (113 °F) in summer and plunge below freezing point at night during winter.

Such large temperature variations have a destructive effect on the exposed rocky surfaces. The repeated fluctuations put a strain on exposed rock and the flanks of mountains crack and shatter. Fragmented strata slide down into the valleys where they continue to break into pieces due to the relentless sun by day and chill by night. Successive strata are exposed to further weathering. The relief of the internal pressure that has built up in rocks that have been underground for aeons can cause them to shatter.

Exfoliation also occurs when the outer surfaces of rocks split off in flat flakes. This is believed to be caused by the stresses put on the rock by repeated thermal expansions and contractions which induces fracturing parallel to the original surface.

Chemical weathering processes probably play a more important role in deserts than was previously thought. The necessary moisture may be present in the form of dew or mist. Ground water may be drawn to the surface by evaporation and the formation of salt crystals may dislodge rock particles as sand or disintegrate rocks by exfoliation. Shallow caves are sometimes formed at the base of cliffs by this means.

As the desert mountains decay, large areas of shattered rock and rubble occur. The process continues and the end products are either dust or sand. Dust is formed from solidified clay or volcanic deposits whereas sand results from the fragmentation of harder granites, limestone and sandstone.

There is a certain critical size (about 0.5 mm) below which further temperature-induced weathering of rocks does not occur and this provides a minimum size for sand grains.

As the mountains are eroded, more and more sand is created. At high wind speeds, sand grains are picked up off the surface and blown along, a process known as saltation. The whirling airborne grains act as a sand blasting mechanism which grinds away solid objects in its path as the kinetic energy of the wind is transferred to the ground.

The sand eventually ends up deposited in level areas known as sand-fields or sand-seas, or piled up in dunes.

Dust storms and sandstorms

Sand and dust storms are natural events that occur in arid regions where the land is not protected by a covering of vegetation. Dust storms usually start in desert margins rather than the deserts themselves where the finer materials have already been blown away. As a steady wind begins to blow, fine particles lying on the exposed ground begin to vibrate. At greater wind speeds, some particles are lifted into the air stream. When they land, they strike other particles which may be jerked into the air in their turn, starting a chain reaction. Once ejected, these particles move in one of three possible ways, depending on their size, shape and density; suspension, saltation or creep. Suspension is only possible for particles less than 0.1 mm (0.0039 in) in diameter. In a dust storm, these fine particles are lifted up and wafted aloft to heights of up to 6 km (3.7 mi). They reduce visibility and can remain in the atmosphere for days on end, conveyed by the trade winds for distances of up to 6,000 km (3,700 mi).

Denser clouds of dust can be formed in stronger winds, moving across the land with a billowing leading edge. The sunlight can be obliterated and it may become as dark as night at ground level.

In a study of a dust storm in China in 2001, it was estimated that 6.5 million tons of dust were involved, covering an area of 134,000,000 km2 (52,000,000 sq mi). The mean particle size was 1.44 μm.

A much smaller scale, short-lived phenomenon can occur in calm conditions when hot air near the ground rises quickly through a small pocket of cooler, low-pressure air above forming a whirling column of particles, a dust devil.

Sandstorms occur with much less frequency than dust storms. They are often preceded by severe dust storms and occur when the wind velocity increases to a point where it can lift heavier particles. These grains of sand, up to about 0.5 mm (0.020 in) in diameter are jerked into the air but soon fall back to earth, ejecting other particles in the process. Their weight prevents them from being airborne for long and most only travel a distance of a few meters (yards). The sand streams along above the surface of the ground like a fluid, often rising to heights of about 30 cm (12 in).

In a really severe steady blow, 2 m (6 ft 7 in) is about as high as the sand stream can rise as the largest sand grains do not become airborne at all. They are transported by creep, being rolled along the desert floor or performing short jumps.

During a sandstorm, the wind-blown sand particles become electrically charged. Such electric fields, which range in size up to 80 kV/m, can produce sparks and cause interference with telecommunications equipment. They are also unpleasant for humans and can cause headaches and nausea.

The electric fields are caused by the collision between airborne particles and by the impacts of saltating sand grains landing on the ground. The mechanism is little understood but the particles usually have a negative charge when their diameter is under 250 µm and a positive one when they are over 500 µm.

Major deserts

Deserts take up about one third of Earth's land surface.

Bottomlands may be salt-covered flats. Eolian processes are major factors in shaping desert landscapes. Polar deserts (also seen as "cold deserts") have similar features, except the main form of precipitation is snow rather than rain. Antarctica is the world's largest cold desert (composed of about 98% thick continental ice sheet and 2% barren rock). Some of the barren rock is to be found in the so-called Dry Valleys of Antarctica that almost never get snow, which can have ice-encrusted saline lakes that suggest evaporation far greater than the rare snowfall due to the strong katabatic winds that even evaporate ice.

Solar Energy Resources

Deserts are increasingly seen as sources for solar energy, partly due to lower cloud cover. Many successful solar power plants have been built in the Mojave Desert. These plants have a combined capacity of 354 megawatts(MW) making them the largest solar power installation in the world. Large swaths of the desert are covered in mirrors (used for solar energy), including nine fields of solar collectors. The Mojave Solar Park is currently under construction and will produce 280MW when completed.

The potential of generating solar energy from the Sahara desert is immense. Professor David Faiman of Ben-Gurion University has stated that the technology now exists to supply all of the world's electricity needs with 10% of the Sahara desert. Desertec Industrial Initiative is a consortium seeking $560 billion investment in North African solar and wind installations over the next 40 years to supply electricity to Europe via cable lines running under the Mediterranean Sea.

European interest in the Sahara desert stems from its two aspects: amount of sunshine and empty space. The Sahara receives more sunshine per are than the sunniest of regions in Europe. The Sahara desert also has the empty space required to house fields of mirrors for solar plants, totalling hundreds of square miles.

The Negev Desert, Israel, and the surrounding area, including the Arava Valley, receive plenty of sunshine and are generally not arable. This has resulted in the construction of many solar plants. David Faiman has proposed that "giant" solar plants in the Negev could supply all of Israel's electricity.

Water Crisis

A general term used to describe a situation where the available water within a region is less than the region's demand. The term has been used to describe the availability of potable water in a variety of regions by the United Nations and other world organizations. Others, for example the Food and Agriculture Organization, said in 2003 that there is no water crisis but steps must be taken to avoid one in the future. The major aspects of the water crisis are allegedly overall scarcity of usable water and water pollution.

According to Nature (2010), about 80% of the world's population (5.6 billion in 2011) live in areas with threats to water security. The water security is a shared threat to human and nature and it is pandemic. Human water-management strategies can affect detrimentally to wildlife, such as migrating fish. Regions with intensive agriculture and dense population, as the US and Europe, have high threat to water security. The researcher estimate that during 2010-2015, ca US$800 billion will be required to cover the annual global investment in water infrastructure. Good management of water resources can jointly manage biodiversity protection and human water security. Preserving flood plains rather than constructing flood-control reservoirs would provide a cost-effective way to control floods while protecting the biodiversity of wildlife that occupies such areas.

The New York Times article, "Southeast Drought Study Ties Water Shortage to Population, Not Global Warming", summarizes the findings of Columbia University researcher on the subject of the droughts in the southwest between 2005 and 2007. The findings were published in the Journal of Climate. They say the water shortages resulted from population size more than rainfall. Census figures show that Georgia's population rose to 9.54 million from 6.48 million between 1990 and 2007.

After studying data from weather instruments, computer models and measurements of tree rings which reflect rainfall, they found that the droughts were not unprecedented and result from normal climate patterns and random weather events. "Similar droughts unfolded over the last thousand years, the researchers wrote. Regardless of climate change, they added, similar weather patterns can be expected regularly in the future, with similar results." As the temperature increases, rainfall in the Southeast will increase but because of evaporation the area may get even drier. The researchers concluded with a statement saying that any rainfall comes from complicated internal processes in the atmosphere that are very hard to predict because of the large amount of variables.

Lawrence Smith, the president of the population institute, asserts that although an overwhelming majority of the planet is composed of water, 97% of this water is constituted of saltwater; the fresh water used to sustain humans is only 3% of the total amount of water on Earth (Hoevel). Therefore, Smith believes that the competition for water in an overpopulated world would pose a major threat to human stability (Hoevel); indeed, world wars may be fought over the control of thinning ice sheets and nearly desiccated reservoirs. 2 billion people have gained access to a safe water source since 1990.

The proportion of people in developing countries with access to safe water is calculated to have improved from 30 percent in 1970 to 71 percent in 1990, 79 percent in 2000 and 84 percent in 2004, parallel with rising population. This trend is projected to continue. The Earth has a limited supply of fresh water, stored in aquifers, surface waters and the atmosphere. Sometimes oceans are mistaken for available water, but the amount of energy needed to convert saline water to potable water is prohibitive today, explaining why only a very small fraction of the world's water supply derives from desalination.

Manifestations

- There are several principal manifestations of the water crisis.
- Inadequate access to safe drinking water for about 884 million people
- Inadequate access to water for sanitation and waste disposal for 2.5 billion people
- Groundwater overdrafting (excessive use) leading to diminished agricultural yields
- Overuse and pollution of water resources harming biodiversity
- Regional conflicts over scarce water resources sometimes resulting in warfare

Waterborne diseases and the absence of sanitary domestic water are one of the leading causes of death worldwide. For children under age five, waterborne diseases are the leading cause of death. At any given time, half of the world's hospital beds are occupied by patients suffering from waterborne diseases. According to the World Bank, 88 percent of all waterborne diseases are caused by unsafe drinking water, inadequate sanitation and poor hygiene.

Water is the underlying tenuous balance of safe water supply, but controllable factors such as the management and distribution of the water supply itself contribute to further scarcity.

A 2006 United Nations report focuses on issues of governance as the core of the water crisis, saying "There is enough water for everyone" and "Water insufficiency is often due to mismanagement, corruption, lack of appropriate institutions, bureaucratic inertia and a shortage of investment in both human capacity and physical infrastructure". Official data also shows a clear correlation between access to safe water and GDP per capita.

It has also been claimed, primarily by economists, that the water situation has occurred because of a lack of property rights, government regulations and subsidies in the water sector, causing prices to be too low and consumption too high.

Vegetation and wildlife are fundamentally dependent upon adequate freshwater resources. Marshes, bogs and riparian zones are more obviously dependent upon sustainable water supply, but forests and other upland ecosystems are equally at risk of significant productivity changes as water availability is diminished. In the case of wetlands, considerable area has been simply taken from wildlife use to feed and house the expanding human population. But other areas have suffered reduced productivity from gradual diminishing of freshwater inflow, as upstream sources are diverted for human use. In seven states of the U.S. over 80 percent of all historic wetlands were filled by the 1980s, when Congress acted to create a "no net loss" of wetlands.

In Europe extensive loss of wetlands has also occurred with resulting loss of biodiversity. For example many bogs in Scotland have been developed or diminished through human population expansion. One example is the Portlethen Moss in Aberdeenshire.

On Madagascar's highland plateau, a massive transformation occurred that eliminated virtually all the heavily forested vegetation in the period 1970 to 2000. The slash and burn agriculture eliminated about ten percent of the total country's native biomass and converted it to a barren wasteland. These effects were from overpopulation and the necessity to feed poor indigenous peoples, but the adverse effects included widespread gully erosion that in turn produced heavily silted rivers that "run red" decades after the deforestation. This eliminated a large amount of usable fresh water and also destroyed much of the riverine ecosystems of several large west-flowing rivers. Several fish species have been driven to the edge of extinction and some, such as, the disturbed Tokios, coral reef formations in the Indian Ocean are effectively lost.

In October 2008, Peter Brabeck-Letmathe, chairman and former chief executive of Nestlé, warned that the production of biofuels will further deplete the world's water supply.

There are many other countries of the world that are severely impacted with regard to human health and inadequate drinking water. The following is a partial list of some of the countries with significant populations (numerical population of affected population listed) whose only consumption is of contaminated water:

- Sudan 12.3 million
- Venezuela 5.0 million
- Ethiopia 2.7 million
- Tunisia 2.1 million
- Cuba 1.3 million

According to the California Department of Water Resources, if more supplies aren't found by 2020, the region will face a shortfall nearly as great as the amount consumed today. Los Angeles is a coastal desert able to support at most 1 million people on its own water; the Los Angeles basin now is the core of a megacity that spans 220 miles (350 km) from Santa Barbara to the Mexican border. The region's population is expected to reach 41 million by 2020, up from 28 million in 2009. The population of California continues to grow by more than two million a year and is expected to reach 75 million in 2030, up from 49 million in 2009. But water shortage is likely to surface well before then.

Water deficits, which are already spurring heavy grain imports in numerous smaller countries, may soon do the same in larger countries, such as China and India. The water tables are falling in scores of countries (including Northern China, the US, and India) due to widespread overpumping using powerful diesel and electric pumps. Other countries affected includePakistan, Iran, and Mexico. This will eventually lead to water scarcity and cutbacks in grain harvest. Even with the overpumping of its aquifers, China is developing a grain deficit. When this happens, it will almost certainly drive grain prices upward. Most of the 3 billion people projected to be added worldwide by mid-century will be born in countries already experiencing water shortages. Unless population growth can be slowed quickly it is feared that there may not be a practical non-violent or humane solution to the emerging world water shortage.

After China and India, there is a second tier of smaller countries with large water deficits — Algeria, Egypt, Iran, Mexico, and Pakistan. Four of these already import a large share of their grain. But with a population expanding by 4 million a year, it will also likely soon turn to the world market for grain.

According to a UN climate report, the Himalayan glaciers that are the sources of Asia's biggest rivers - Ganges, Indus, Brahmaputra, Yangtze, Mekong, Salween and Yellow - could disappear by 2035 as temperatures rise. It was later revealed that the source used by the UN climate report actually stated 2350, not 2035. Approximately 2.4 billion people live in the drainage basin of the Himalayan rivers. India, China, Pakistan, Bangladesh, Nepal and Myanmar could experience floods followed by droughts in coming decades. In India alone, the Ganges provides water for drinking and farming for more than 500 million people. The west coast of North America, which gets much of its water from glaciers in mountain ranges such as the Rocky Mountains and Sierra Nevada, also would be affected.

By far the largest part of Australia is desert or semi-arid lands commonly known as the outback. In June 2008 it became known that an expert panel had warned of long term, possibly irreversible, severe ecological damage for the whole Murray-Darling basin if it does not receive sufficient water by October. Water restrictions are currently in place in many regions and cities of Australia in response to chronic shortages resulting from drought.

The Australian of the year 2007, environmentalist Tim Flannery, predicted that unless it made drastic changes, Perth in Western Australia could become the world's first ghost metropolis, an abandoned city with no more water to sustain its population. However, Western Australia's dams reached 50% capacity for the first time since 2000 as of September 2009. As a result, heavy rains have brought forth positive results for the region. Nonetheless, the following year, 2010, Perth suffered its second-driest winter on record and the water corporation tightened water restrictions for spring.

Effects on Climate

Aquifer drawdown or overdrafting and the pumping of fossil water increases the total amount of water within the hydrosphere subject to transpiration and evaporation processes, thereby causing accretion in water vapour and cloud cover, the primary absorbers of infrared radiation in the earth's atmosphere. Adding water to the system has a forcing effect on the whole earth system, an accurate estimate of which hydrogeological fact is yet to be quantified.

Construction of wastewater treatment plants and reduction of groundwater overdrafting appear to be obvious solutions to the worldwide problem; however, a deeper look reveals more fundamental issues in play. Wastewater treatment is highly capital intensive, restricting access to this technology in some regions; furthermore the rapid increase in population of many countries makes this a race that is difficult to win. As if those factors are not daunting enough, one must consider the enormous costs and skill sets involved to maintain wastewater treatment plants even if they are successfully developed.

Reduction in groundwater overdrafting is usually politically very unpopular and has major economic impacts to farmers; moreover, this strategy will necessarily reduce crop output, which is something the world can ill-afford, given the population level at present.

At more realistic levels, developing countries can strive to achieve primary wastewater treatment or secure septic systems, and carefully analyse wastewater outfall design to minimise impacts to drinking water and to ecosystems. Developed countries can not only share technology better, including cost-effective wastewater and water treatment systems but also in hydrological transport modeling. At the individual level, people in developed countries can look inward and reduce overconsumption, which further strains worldwide water consumption. Both developed and developing countries can increase protection of ecosystems, especially wetlands and riparian zones. These measures will not only conserve biota, but also render more effective the natural water cycle flushing and transport that make water systems more healthy for humans.

A range of local, low-tech solutions are being pursued by a number of companies. These efforts center around the use of solar power to distill water at temperatures slightly beneath that at which water boils. By developing the capability to purify any available water source, local business models could be built around the new technologies, accelerating their uptake.

Conventional Fossil or Nuclear Energy Based Desalination

As new technological innovations continue to reduce the capital cost of desalination, more countries are building desalination plants as a small element in addressing their water crises.

- Israel desalinizes water for a cost of 53 cents per cubic meter
- Singapore desalinizes water for 49 cents per cubic meter and also treats sewage with reverse osmosis for industrial and potable use (NEWater).
- China and India, the world's two most populous countries, are turning to desalination to provide a small part of their water needs

- In 2007 Pakistan announced plans to use desalination
- All Australian capital cities (except Darwin, Northern Territory and Hobart) are either in the process of building desalination plants, or are already using them. In late 2011, Melbournewill begin using Australia's largest desalination plant, the Wonthaggi desalination plant to raise low reservoir levels.
- In 2007 Bermuda signed a contract to purchase a desalination plant
- The largest desalination plant in the United States is the one at Tampa Bay, Florida, which began desalinizing 25 million gallons (95000 m^3) of water per day in December 2007. In the United States, the cost of desalination is $3.06 for 1,000 gallons, or 81 cents per cubic meter. In the United States, California, Arizona, Texas, and Florida use desalination for a very small part of their water supply.
- After being desalinized at Jubail, Saudi Arabia, water is pumped 200 miles (320 km) inland though a pipeline to the capital city of Riyadh.

A January 17, 2008, article in the Wall Street Journal states, "World-wide, 13,080 desalination plants produce more than 12 billion gallons of water a day, according to the International Desalination Association."

The world's largest desalination plant is the Jebel Ali Desalination Plant (Phase 2) in the United Arab Emirates. It is a dual-purpose facility that uses multi-stage flash distillation and is capable of producing 300 million cubic meters of water per year.

A typical aircraft carrier in the U.S. military uses nuclear power to desalinize 400,000 US gallons (1,500,000 L) of water per day. While desalinizing 1,000 US gallons (3,800 L) of water can cost as much as $3, the same amount of bottled water costs $7,945.

However, given the energy intensive nature of desalination, with associated economic and environmental costs, desalination is generally considered a last resort after water conservation. But this is changing as prices continue to fall.

According to MSNBC, a report by Lux Research estimated that the worldwide desalinated water supply will triple between 2008 and 2020. However, not everyone is convinced that desalination is or will be economically viable or environmentally sustainable for the foreseeable future.Debbie Cook, the former mayor of Huntington Beach, California, has been a frequent critic of desalination proposals ever since she was appointed as a member of the California Desalination Task Force.

Cook claims that in addition to being energy intensive, desalination schemes are very costly—often much more costly than desalination proponents claim. In her writing on the subject, Cook points to a long list of projects that have stalled or been aborted for financial or other reasons, and suggests that water-stressed regions would do better to focus on conservation or other water supply solutions than to invest in desalination plants.

Solar Energy Based Desalination

A novel approach to desalination is the Seawater Greenhouse which takes seawater and uses solar energy to desalinate it in conjunction with growing food crops in a specially adapted greenhouse.

Global Experiences in Managing water Crisis

It is alleged that the likelihood of conflict rises if the rate of change within the basin exceeds the capacity of institution to absorb that change. Although water crisis is closely related to regional tensions, history showed that acute conflicts over water are far less than the record of cooperation.

The key lies in strong institutions and cooperation. The Indus River Commission and the Indus Water Treaty survived two wars between India and Pakistan despite their hostility, proving to be a successful mechanism in resolving conflicts by providing a framework for consultation inspection and exchange of data. The Mekong Committee has also functioned since 1957 and survived the Vietnam War. In contrast, regional instability results when there is an absence of institutions to co-operate in regional collaboration, like Egypt's plan for a high dam on the Nile.

However, there is currently no global institution in place for the management and management of trans-boundary water sources, and international co-operation has happened through ad hoc collaborations between agencies, like the Mekong Committee which was formed due to an alliance between UNICEF and the US Bureau of Reclamation. Formation of strong international institutions seems to be a way forward - they fuel early intervention and management, preventing the costly dispute resolution process.

One common feature of almost all resolved disputes is that the negotiations had a "need-based" instead of a "right–based" paradigm. Irrigable lands, population, technicalities of projects define "needs". The success of a need-based paradigm is reflected in the only water agreement ever negotiated in the Jordan River Basin, which focuses in needs not on rights of riparians. In the Indian subcontinent, irrigation requirements of Bangladesh determine water allocations of The Ganges River. A need based, regional approach focuses on satisfying individuals with their need of water, ensuring that minimum quantitative needs are being met. It removes the conflict that arises when countries view the treaty from a national interest point of view, move away from the zero-sum approach to a positive sum, integrative approach that equitably allocated the water and its benefits.

Human Life in Deserts

A desert is a hostile, potentially deadly environment for unprepared humans. In hot deserts, high temperatures cause rapid loss of water due to sweating, and the absence of water sources with which to replenish it can result in dehydration and death within a few days. In addition, unprotected humans are also at risk from heatstroke.

Humans may also have to adapt to sandstorms in some deserts, not just in their adverse effects on respiratory systems and eyes, but also in their potentially harmful effects on equipment such as filters, vehicles and communication equipment. Sandstorms can last for hours, sometimes even days. This makes surviving in the desert quite difficult for humans.

Despite this, some cultures have made hot deserts their home for thousands of years, including the Bedouin, Tuareg and Pueblo people. Modern technology, including advanced irrigation systems, desalinization and air conditioning have made deserts much more hospitable. In theUnited States and Australia for example, desert farming has found extensive use.

In cold deserts, hypothermia and frostbite are the chief hazards, as well as dehydration in the absence of a source of heat to melt ice for drinking. Falling through pack-ice or surface ice layers into freezing water is a particular danger requiring emergency action to prevent rapid hypothermia. Starvation is also a hazard; in low temperatures the body requires much more food energy to maintain body heat and to move. As with hot deserts, some people such as the Inuit have adapted to the harsh conditions of cold deserts.

Most traditional human life in deserts is nomadic. It depends in hot deserts on finding water, and on following infrequent rains to obtain grazing for livestock. In cold deserts, it depends on finding good hunting and fishing grounds, on sheltering from blizzards and winter extremes, and on storing enough food for winter. Permanent settlement in both kinds of deserts requires permanent water and food sources and adequate shelter, or the technology and energy sources to provide it.

Many deserts are flat and featureless, lacking landmarks, or composed of repeating landforms such as sand dunes or the jumbled ice-fields of glaciers. Advanced skills or devices are required to navigate through such landscapes and inexperienced travellers may perish when supplies run out after becoming lost. In addition sandstorms or blizzards may cause disorientation in severely reduced visibility.

The danger represented by wild animals in deserts has been featured in explorers' accounts but does not cause higher rates of death than in other environments such as rain forests or savanna woodland, and generally does not by itself affect human distribution. Defense against polar bears may be advisable in some areas of the Arctic, as may precautions against venomous snakes and scorpions in choosing sites at which to camp in some hot deserts.

Desert Greening

Desert greening consists of any of a number of methods used to revitalize deserts. So far only arid and semi-arid desert are meant when using the expression. The icy-deserts and others are considered unsuitable.

Methods

- Landscaping methods to reduce evaporation, erosion, consolidation of topsoil, sandstorms, temperature and more
- Permaculture in general - growing of plant communities Polyculture, composting or multi trophic agriculture
- Planting trees (Pioneer species) and salt-loving plants such as Salicornia and Halophytes
- Regeneration of salty, polluted, or degenerated soils
- Floodwater retention and infiltration (Flood control)

- Greenhouse agriculture like the Integrated biotectural system
- Seawater farming like done by the Seawater foundation
- Inland Mariculture
- Prevention of overgrazing and firewood use
- Training of local residents to care for plantings, water systems etc.

Water

Desert greening is more or less a function of water availability. If sufficient water for irrigation is at hand any hot, cold, sandy or rocky desert can be greened. Water can be made available through saving, reuse, rainwater harvesting, desalination, or direct use of seawater for salt-loving plants. These different paths have unique features, each: Saving water is for free - reuse of treated water and the closing of cycles is the most promising because closed cycles stand for unlimited and sustainable supply - rainwater management is a decentralized solution and applicable for inland areas - desalination is very secure as long as the primary energy for the operation of the desalination plant is available - Direct use of seawater for seawater agriculture is the most potent, only limited by the need for pumping up the water from sea-level.

There are theoretical water sources, too. Like the generation of artificial rain through cloud seeding of various kinds. An available technology called "atmospheric water generation" or air to water is used by the military and is available as micro-solution for drinking water fountains. But this technology uses 200 times more energy than modern desalination plants and cannot be considered for desert-greening.

Countering Desertification

The soil of the Thar Desert in India remains dry for much of the year and is prone to wind erosion. High velocity winds blow soil from the desert, depositing some on neighboring fertile lands, and causing shifting sand dunes within the desert, which buries fences and block roads and railway tracks. Permanent solution to this problem of shifting sand dunes can be provided by planting appropriate species on the dunes to prevent further shifting and planting windbreaks and shelterbelts. These solutions also provide protection from hot or cold and desiccating winds and the invasion of sand. The Rajasthan Canal system is the major irrigation scheme of the Thar Desert and is intended to reclaim it and to check spreading of the desert to fertile areas.

Prevention of shifting sand dunes is accomplished through plantations of Acacia tortilis near Laxmangarh town. There are few local tree species suitable for planting in the desert region and these are slow growing. The introduction of exotic tree species in the desert for plantation has become necessary. Many species of Eucalyptus, Acacia, Cassia and other genera from Israel, Australia, US, Russia, Zimbabwe, Chile, Peru, and Sudan have been tried in the Thar Desert. Acacia tortilis has proved to be the most promising species for desert greening and. The jojoba is another promising species of economic value which has been found suitable for planting in these areas.

History

The world's great deserts were formed by natural processes interacting over long intervals of time. During most of these times, deserts have grown and shrunk independent of human activities. Paleodeserts are large sand seas now inactive because they are stabilized by vegetation, some extending beyond the present margins of core deserts, such as the Sahara, the largest desert.

Desertification has played a significant role in human history, contributing to the collapse of several large empires, such as Carthage, Greece, and the Roman Empire, as well as causing displacement of local populations.

Areas Affected

Drylands occupy approximately 40-41% of Earth's land area and are home to more than 2 billion people. It has been estimated that some 10–20% of drylands are already degraded, the total area affected by desertification being between 6 and 12 million square kilometres, that about 1–6% of the inhabitants of drylands live in desertified areas, and that a billion people are under threat from further desertification. The Sahara is currently expanding south at a rate of up to 48 kilometres per year.

Causes

Dryland ecosystems are already very fragile, and can rarely sustain the increased pressures that result from intense population growth. Many of these areas are inappropriately opened to development, when they cannot sustain human settlements.

The most common cause of desertification is the over cultivation of desert lands. Over-cultivation causes the nutrients in the soil to be depleted faster than they are restored. Improper irrigation practices result in salinated soils, and depletion of aquifers.

Vegetation plays a major role in determining the biological composition of the soil. Studies have shown that, in many environments, the rate of erosion and runoff decreases exponentially with increased vegetation cover. Overgrazing removes this vegetation causing erosion and loss of topsoil.

Desertification and poverty

At least 90% of the inhabitants of drylands live in developing nations, where they also suffer from poor economic and social conditions. This situation is exacerbated by land degradation because of the reduction in productivity, the precariousness of living conditions and the difficulty of access to resources and opportunities.

A downward spiral is created in many underdeveloped countries by overgrazing, land exhaustion and overdrafting of groundwater in many of the marginally productive world regions due to overpopulation pressures to exploit marginal drylands for farming. Decision-makers are understandably averse to invest in arid zones with low potential. This absence of investment contributes to the marginalisation of these zones. When unfavourable agro-climatic conditions are combined with an absence of infrastructure and access to markets, as well as poorly adapted production techniques and an underfed and undereducated population, most such zones are excluded from development.

Desertification often causes rural lands to become unable to support the same sized populations that previously lived there. This results in mass migrations out of rural areas and into urban areas, particularly in Africa. Because of these migrations into the cities, there are often large numbers of unemployed people who end up living in slums.

Countermeasures and Prevention

Techniques exist for mitigating or reversing the effects of desertification, however there are numerous barriers to their implementation. One of these is that the costs of adopting sustainable agricultural practices sometimes exceed the benefits for individual farmers, even while they are socially and environmentally beneficial. Another issue is a lack of political will, and lack of funding to support land reclamation and anti-desertification programs.

Desertification is recognized as a major threat to biodiversity. Some countries have developed Biodiversity Action Plans to counter its effects, particularly in relation to the protection of endangered flora and fauna.

Reforestation gets at one of the root causes of desertification and isn't just a treatment of the symptoms. Environmental organizations work in places where deforestation and desertification are contributing to extreme poverty. There they focus primarily on educating the local population about the dangers of deforestation and sometimes employ them to grow seedlings, which they transfer to severely deforested areas during the rainy season.

Techniques focus on two aspects: provisioning of water, and fixation and hyper-fertilizing soil.

Fixating the soil is often done through the use of shelter belts, woodlots and windbreaks. Windbreaks are made from trees and bushes and are used to reduce soil erosion and evapotranspiration. They were widely encouraged by development agencies from the middle of the 1980s in the Sahel area of Africa.

Some soils (for example, clay), due to lack of water can become consolidated rather than porous (as in the case of sandy soils). Some techniques as zaï or tillage are then used to still allow the planting of crops.

Enriching of the soil and restoration of its fertility is often done by plants. Of these, the Leguminous plants which extract nitrogen from the air and fixes it in the soil, and food crops/trees as grains, barley, beans and dates are the most important. Sand fences can also be used to control drifting of soil and sand erosion.

As there are many different types of deserts, there are also different types of desert reclamation methodologies. An example for this is the salt-flats in the Rub' al Khali desert in Saudi-Arabia. These salt-flats are one of the most promising desert areas for seawater agriculture and could be revitalized without the use of freshwater or much energy.

Farmer Managed Natural Regeneration (FMNR) is another technique that has produced successful results for desert reclamation. Since 1980, this method to reforest degraded landscape has been applied with some success in Niger. This simple and low-cost method has enabled farmers to regenerate some 30,000 square kilometers in Niger. The process involves enabling native sprouting tree growth through selective pruning of shrub shoots.

The residue from pruned tress can be used to provide mulching for fields thusincreasing soil water retention and reducing evaporation. Additionally, properly spaces and pruned trees can increase crop yields. The Humbo Assisted Regeneration Project which uses FMNR techniques in Ethiopia has received money from The World Bank's BioCarbon Fund, which supports projects that sequester or conserve carbon in forests or agricultural ecosystems.

Dune

In physical geography, a dune is a hill of sand built by wind. Dunes occur in different forms and sizes, formed by interaction with the wind. Most kinds of dunes are longer on the windward side where the sand is pushed up the dune and have a shorter "slip face" in the lee of the wind. The valley or trough between dunes is called a slack. A "dune field" is an area covered by extensive sand dunes. Large dune fields are known as ergs.

Some coastal areas have one or more sets of dunes running parallel to the shoreline directly inland from the beach. In most cases the dunes are important in protecting the land against potential ravages by storm waves from the sea. Although the most widely distributed dunes are those associated with coastal regions, the largest complexes of dunes are found inland in dry regions and associated with ancient lake or sea beds.

Dunes also form under the action of water flow (alluvial processes), and on sand or gravel beds of rivers, estuaries and the sea-bed.

The modern word "dune" came into English from French circa 1790. In ancient times, words cognate to "dune" probably had the meaning of a built-up hill or citadel fortification.

Conservation

Dune habitats provide niches for highly specialised plants and animals, including numerous rare species and some endangered species. Due to widespread human population expansion, dunes face destruction through land development and recreational usages, as well as alteration to prevent the encroachment of sand onto inhabited areas. Some countries, notably the United States, Australia, Canada, New Zealand, the United Kingdom, and Netherlands, have developed significant programs of dune protection through the use of sand dune stabilization. In the U.K., a Biodiversity Action Plan has been developed to assess dunes loss and to prevent future dunes destruction.

Aeolian dune shapes

Crescentic

Crescent-shaped mounds are generally wider than they are long. The slipfaces are on the concave sides of the dunes. These dunes form under winds that blow consistently from one direction, and they also are known as barchans, or transverse dunes. Some types of crescentic dunes move more quickly over desert surfaces than any other type of dune. A group of dunes moved more than 100 metres per year between 1954 and 1959 in the China's Ningxia Province, and similar speeds have been recorded in the Western Desert of Egypt. The largest crescentic dunes on Earth, with mean crest-to-crest widths of more than 3 kilometres, are in China's Taklamakan Desert.

Linear

Straight or slightly sinuous sand ridges typically much longer than they are wide are known as linear dunes. They may be more than 160 kilometres (99 mi) long. Some linear dunes merge to form Y-shaped compound dunes. Many form in bidirectional wind regimes. The long axes of these dunes extend in the resultant direction of sand movement. Linear loess hills known as pahas are superficially similar. These hills appear to have been formed during the last age under permafrost conditions dominated by sparse tundra vegetation.

Star

Radially symmetrical, star dunes are pyramidal sand mounds with slipfaces on three or more arms that radiate from the high center of the mound. They tend to accumulate in areas with multidirectional wind regimes. Star dunes grow upward rather than laterally. They dominate the Grand Erg Oriental of the Sahara. In other deserts, they occur around the margins of the sand seas, particularly near topographic barriers. In the southeast Badain Jaran Desert of China, the star dunes are up to 500 metres tall and may be the tallest dunes on Earth.

Dome

Oval or circular mounds that generally lack a slipface, dome dunes are rare, and these occur at the far upwind margins of sand seas.

Parabolic

U-shaped mounds of sand with convex noses trailed by elongated arms are parabolic dunes. These dunes are formed from blowout dunes where the erosion of vegetated sand leads to a U-shaped depression. The elongated arms are held in place by vegetation; the largest arm known on Earth reaches 12 km. Sometimes these dunes are called U-shaped, blowout, or hairpin dunes, and they are well known in coastal deserts. Unlike crescent shaped dunes, their crests point upwind. The bulk of the sand in the dune migrates forward.

In plan view, these are U-shaped or V-shaped mounds of well-sorted, very fine to medium sand with elongated arms that extend upwind behind the central part of the dune. There are slip faces that often occur on the outer side of the nose and on the outer slopes of the arms.

These dunes often occur in semiarid areas where the precipitation is retained in the lower parts of the dune and underlying soils. Parabolic dunes are dependent on the vegetation that covers them—grasses,shrubs,and trees, which help anchor the trailing arms. In inland deserts, parabolic dunes commonly originate and extend downwind from blowouts in sand sheets only partly anchored by vegetation. They can also originate from beach sands and extend inland into vegetated areas in coastal zones and on shores of large lakes.

Most parabolic dunes do not reach heights higher than a few tens of metres except at their nose, where vegetation stops or slows the advance of accumulating sand.

Simple parabolic dunes have only one set of arms that trail upwind, behind the leading nose. Compound parabolic dunes are coalesced features with several sets of trailing arms. Complex parabolic dunes include subsidiary superposed or coalesced forms, usually of barchanoid or linear shapes.

Parabolic dunes, like crescent dunes, occur in areas where very strong winds are mostly unidirectional. Although these dunes are found in areas now characterized by variable wind speeds, the effective winds associated with the growth and migration of both the parabolic and crescent dunes probably are the most consistent in wind direction.

The grain size for these well-sorted, very fine to medium sands is about 0.06 to 0.5 mm. Parabolic dunes have loose sand and steep slopes only on their outer flanks. The inner slopes are mostly well packed and anchored by vegetation, as are the corridors between individual dunes. Because all dune arms are oriented in the same direction, and, the inter-dune corridors are generally swept clear of loose sand, the corridors can usually be traversed in between the trailing arms of the dune. However to cross straight over the dune by going over the trailing arms, can be very difficult. Also, traversing the nose is very difficult as well because the nose is usually made up of loose sand without much if any vegetation.

Longitudinal (Seif) Dunes

Longitudinal dunes (also called Seif dunes, after the Arabic word for "sword"), elongate parallel to the prevailing wind, possibly caused by a larger dune having its smaller sides blown away. Seif dunes are sharp-crested and are common in the Sahara. They range up to 300 m (980 ft) in height and 300 km (190 mi) in length. In the southern third of the Arabian Peninsula, a vast erg called the Rub' al Khali or the Empty Quarter, contains seif dunes that stretch for almost 200 km and reach heights of over 300 m.

Seif dunes are thought to develop from barchans if a change of the usual wind direction occurs. The new wind direction will lead to the development of a new wing and the over development of one of the original wings. If the prevailing wind then becomes dominant for a lengthy period of time the dune will revert to its barchan form, with one exaggerated wing. Should the strong wind then return the exaggerated wing will further extend so that eventually it will be supplied with sand when the prevailing wind returns. The wing will continue to grow under both wind conditions, thus producing a seif dune. On a seif dune the slip face develops on the side facing away from the strong wind, while the slip face of a barchan faces the direction of movement. In the sheltered troughs between highly developed seif dunes barchans may be formed because the wind is unidirectional.

A transverse dune is perpendicular to the prevailing wind, probably caused by a steady build-up of sand on an already existing minuscule mound.

Reversing Dunes

Occurring wherever winds periodically reverse direction, reversing dunes are varieties of any of the above shapes. These dunes typically have major and minor slipfaces oriented in opposite directions.

All these dune shapes may occur in three forms: simple, compound, and complex. Simple dunes are basic forms with a minimum number of slipfaces that define the geometric type. Compound dunes are large dunes on which smaller dunes of similar type and slipface orientation are superimposed, and complex dunes are combinations of two or more dune types. A crescentic dune with a star dune superimposed on its crest is the most common complex dune. Simple dunes represent a wind regime that has not changed in intensity or direction since the formation of the dune, while compound and complex dunes suggest that the intensity and direction of the wind has changed.

Sub-aqueous Dunes

Sub-aqueous (underwater) dunes (also known in geology as megaripples) form on a bed of sand or gravel under the actions of water flow. They are ubiquitous in natural channels such as rivers and estuaries, and also form in engineered canals and pipelines. Dunes move downstream as the upstream slope is eroded and the sediment deposited on the downstream or lee slope in typical bedform construction.

These dunes most often form as a continuous 'train' of dunes, showing remarkable similarity in wavelength and height.

Dunes on the bed of a channel significantly increase flow resistance, their presence and growth playing a major part in river flooding.

Lithified Dunes

A lithified (consolidated) sand dune is a type of sandstone that is formed when a marine or aeolian sand dune becomes compacted and hardened. Once in this form, water passing through the rock can carry and deposit minerals, which can alter the color of the rock. Cross-bedded layers of stacks of lithified dunes can produce the cross-hatching patterns, such as those seen in the Zion National Park in the western United States.

A slang term that is used in the Southwestern States (of the U.S.A.) for those consolidated and hardened sand dunes is "slickrock", a name that was introduced by pioneers of the Old West because their steel-rimmed wagon wheels could not gain traction on the rock.

Coastal Dunes

Dunes form where constructive waves encourage the accumulation of sand, and where prevailing onshore winds blow this sand inland. There need to be obstacles — for example, vegetation, pebbles and so on — to trap the moving sand grains. As the sand grains get trapped they start to accumulate, starting dune formation. The wind then starts to affect the mound of sand by eroding sand particles from the windward side and depositing them on the leeward side. Gradually this action causes the dune to "migrate" inland, as it does so it accumulates more and more sand. Dunes provide privacy and shelter from the wind.

Ecological succession on coastal dunes

As a dune forms, plant succession occurs. The conditions on an embryo dune are harsh, with salt spray from the sea carried on strong winds. The dune is well drained and often dry, and composed of calcium carbonate from seashells. Rotting seaweed, brought in by storm waves adds nutrients to allow pioneer species to colonize the dune. These pioneer species are marram grass, sea wort grass and other sea grasses in the United Kingdom. These plants are well adapted to the harsh conditions of the foredune typically having deep roots which reach thewater table, root nodules that produce nitrogen compounds, and protected stoma, reducingtranspiration. Also, the deep roots bind the sand together, and the dune grows into aforedune as more sand is blown over the grasses. The grasses add nitrogen to the soil, meaning other, less hardy plants can then colonize the dunes. Typically these are heather,heaths and gorses. These too are adapted to the low soil water content and have small, prickly leaves which reduce transpiration. Heather adds humus to the soil and is usually replaced by coniferous trees, which can tolerate low soil pH, caused by the accumulation and decomposition of organic matter with nitrate leaching. Coniferous forests and heathland are common climax communities for sand dune systems.

Young dunes are called yellow dunes and dunes which have high humus content are called grey dunes. Leaching occurs on the dunes, washing humus into the slacks, and the slacks may be much more developed than the exposed tops of the dunes. It is usually in the slacks that more rare species are developed and there is a tendency for the dune slacks soil to be waterlogged and where only marsh plants can survive. These plants would include: creeping willow, cotton grass, yellow ins, reeds, and rushes. As for the species, there is a tendency for natterjack toads to breed here.

Sand dunes can have a negative impact on humans when they encroach on human habitats. Sand dunes move via a few different means, all of them helped along by wind. One way that dunes can move is by saltation, where sand particles skip along the ground like a bouncing ball. When these skipping particles land, they may knock into other particles and cause them to move as well, in a process known as creep. With slightly stronger winds, particles collide in mid-air, causing sheet flows. In a major dust storm, dunes may move tens of metres through such sheet flows. Also as in the case of snow, sand avalanches, falling down the slipface of the dunes - that face away from the winds - also move the dunes forward.

Sand threatens buildings and crops in Africa, the Middle East, and China. Drenching sand dunes with oil stops their migration, but this approach is quite destructive to the dunes' animal habitats and uses a valuable resource. Sand fences might also slow their movement to a crawl, but geologists are still analyzing results for the optimum fence designs. Preventing sand dunes from overwhelming towns, villages, and agricultural areas has become a priority for the United Nations Environment Programme. Planting dunes with vegetation also helps to stabilise them.

10

Geography of Volcanoes and Earthquake

Earthquakes and volcanoes are two examples of sudden movements that generate significant deformation in a short period of time. Any sudden shaking of the ground generated by seismic waves passing through Earth's rocks is referred to as an earthquake whereas a volcano is a crack in the earth's crust that allows lava, volcanic ash, and gases to escape. Liquid magma with dissolved gases rises through cracks in the Earth's crust beneath a volcano.

A volcano is an opening in the Earth's crust that allows magma, hot ash and gases to escape. Volcanoes can look like mountains or small hills, depending on what type they are.

Magma is molten rock - rock that is so hot it has turned into liquid. When magma reaches the surface of the Earth it is called lava and comes out of the volcano as a volcanic eruption, along with gases and ash.

The Royal Geographical Society - There is an average of 25 eruptions per year and approximately 600 million people living on or in the vicinity of volcanoes.'

Volcanic Eruptions

Most volcanic eruptions are caused by tectonic plates moving towards each other, which usually produces violent eruptions. Other volcanoes, such as Mauna Loa in Hawaii are caused by hot spots in the Earth's crust. These do not erupt violently and lava usually flows slowly out of them.

What Causes Volcanoes?

The earth has "plates" and when two plates hit each other, one goes under and becomes "molten" which means it gets so hot, the ground turns into liquid. The molten stuff finds a hole in the ground and comes out the top. All of the continents used to be just one big piece of land, but the land broke up and floated away and created the 7 continents we have now. This was called continental drift. If you look at South America, it looks like it could fit like a puzzle piece into Africa.

Because of the way the ground broke, it created plates. Even though the ground is heavy, the plates continue to move and that can cause earthquakes and volcanoes. When two plates hit each other, one can go under and become"molten" which means it gets so hot, the ground turns into liquid. The molten stuff finds a hole in the ground and comes out the top. When the molten stuff is in the volcano it is called magma. When it explodes, it is called lava.

Volcanoes explode because of the gas buildup. When the gas builds up a lot, the gas and lava, which is hot rocks, blows out the top, it's a lot like the way a soda explodes. The lava can get up to 1600 degrees - a lot hotter than an oven. After the lava cools, it turns into rock. One of the types of rock is obsidian, which is shiny and black. Volcanoes can also occur on other planets. Not all volcanoes explode, some are inactive and some are dormant.

The lava can kill people, but there can be other problems. If there's heavy rains on the ash it creates a thick mud that has trapped people and animals. When all the trees and plants are destroyed by the lava, the animals have nothing to eat.

Lots of volcanoes happen in the "Ring of Fire" which is from California to Asia. There are large tectonic plates in the Ring of Fire and lots of earthquakes happen here too.

Volcanic eruptions can cause great damage and the loss of life and property.

The Word Volcano: The word volcano comes from the Roman god of fire, Vulcan. Vulcan was said to have had a forge (a place to melt and shape iron) on Vulcano, an active volcano on the Lipari Islands in Italy.

Extreme Volcanoes: The largest volcano on Earth is Hawaii's Mauna Loa. Mauna Loa is about 6 miles (10 km) tall from the sea floor to its summit (it rises about 4 km above sea level). It also has the greatest volume of any volcano, 10,200 cubic miles (42,500 cubic kilometers). The most active volcano in the continental USA is Mt. St. Helens (located in westernWashington state).

The largest volcano in our Solar System is perhaps Olympus Mons on the planet Mars. This enormous volcano is 17 miles (27 km) tall and over 320 miles (520 km) across.

Divergent Plate Boundaries

At the mid-oceanic ridges, two tectonic plates diverge from one another. New oceanic crust is being formed by hot molten rock slowly cooling and solidifying. The crust is very thin at mid-oceanic ridges due to the pull of the tectonic plates. The release of pressure due to the thinning of the crust leads to adiabatic expansion, and the partial melting of the mantle causing volcanism and creating new oceanic crust. Most divergent plate boundaries are at the bottom of the oceans, therefore most volcanic activity is submarine, forming new seafloor. Black smokers or deep sea vents are an example of this kind of volcanic activity. Where the mid-oceanic ridge is above sea-level, volcanic islands are formed, for example, Iceland.

Convergent Plate Boundaries

Subduction zones are places where two plates, usually an oceanic plate and a continental plate, collide. In this case, the oceanic plate subducts, or submerges under the continental plate forming a deep ocean trench just offshore. Water released from the subducting plate lowers the melting temperature of the overlying mantle wedge, creating magma. This magma tends to be very viscous due to its high silica content, so often does not reach the surface and cools at depth. When it does reach the surface, a volcano is formed. Typical examples for this kind of volcano are Mount Etna and the volcanoes in the Pacific Ring of Fire.

"Hotspots"

"Hotspots" is the name given to volcanic provinces postulated to be formed by mantle plumes. These are postulated to comprise columns of hot material that rise from the core-mantle boundary. They are suggested to be hot, causing large-volume melting, and to be fixed in space. Because the tectonic plates move across them, each volcano becomes dormant after a while and a new volcano is then formed as the plate shifts over the postulated plume. The Hawaiian Islands have been suggested to have been formed in such a manner, as well as theSnake River Plain, with the Yellowstone Caldera being the part of the North American plate currently above the hot spot. This theory is currently under criticism, however.

Volcanic features

The most common perception of a volcano is of a conical mountain, spewing lava and poisonous gases from a crater at its summit; however, this describes just one of the many types of volcano. The features of volcanoes are much more complicated and their structure and behavior depends on a number of factors. Some volcanoes have rugged peaks formed by lava domes rather than a summit crater while others have landscape features such as massive plateaus. Vents that issue volcanic material (including lava and ash) and gases (mainly steam and magmatic gases) can develop anywhere on the landform and may give rise to smaller cones such as Pu?u ?O?o on a flank of Kilauea in Hawaii. Other types of volcano include cryovolcanoes (or ice volcanoes), particularly on some moons of Jupiter, Saturn, and Neptune; and mud volcanoes, which are formations often not associated with known magmatic activity. Active mud volcanoes tend to involve temperatures much lower than those of igneous volcanoes except when the mud volcano is actually a vent of an igneous volcano.

Fissure vents

Volcanic fissure vents are flat, linear fractures through which lava emerges.

Shield volcanoes

Shield volcanoes, so named for their broad, shield-like profiles, are formed by the eruption of low-viscosity lava that can flow a great distance from a vent. They generally do not explode catastrophically, but are characterized by relatively gentle effusive eruptions. Since low-viscosity magma is typically low in silica, shield volcanoes are more common in oceanic than continental settings. The Hawaiian volcanic chain is a series of shield cones, and they are common in Iceland, as well.

Lava domes

Lava domes are built by slow eruptions of highly viscous lava. They are sometimes formed within the crater of a previous volcanic eruption, as in the case of Mount St. Helens, but can also form independently, as in the case of Lassen Peak. Like stratovolcanoes, they can produce violent, explosive eruptions, but the lava generally does not flow far from the originating vent.

Cryptodomes

Cryptodomes are formed when viscous lava is forced upward causing the surface to bulge. The 1980 eruption of Mount St. Helens was an example; lava beneath the surface of the mountain created an upward bulge, which later collapsed down the north side of the mountain.

Cinder cones

Cinder cones result from eruptions of mostly small pieces of scoria and pyroclastics (both resemble cinders, hence the name of this volcano type) that build up around the vent. These can be relatively short-lived eruptions that produce a cone-shaped hill perhaps 30 to 400 meters (100 to 1,300 ft) high. Most cinder cones erupt only once. Cinder cones may form as flank vents on larger volcanoes, or occur on their own. Parícutin in Mexico and Sunset Crater in Arizona are examples of cinder cones. In New Mexico, Caja del Rio is a volcanic field of over 60 cinder cones.

Based on satellite images, it was suggested that cinder cones might occur on other terrestrial bodies in the Solar system too; on the surface of Mars and the Moon.

Stratovolcanoes (composite volcanoes)

Stratovolcanoes (composite volcanoes) are tall conical mountains composed of lava flows and tephra in alternate layers, the strata that gives rise to the name. They are also known as composite volcanoes because they are created from multiple structures during different kinds of eruptions. Classic examples include Mount Fuji in Japan, Mayon Volcano in the Philippines, and Mount Vesuvius and Stromboli in Italy.

Ash produced by the explosive eruption of stratovolcanoes has historically posed the greatest volcanic hazard to civilizations. The lavas of stratovolcanoes are higher in silica, and therefore much more viscous, than lavas from shield volcanoes. High-silica lavas also tend to contain more dissolved gas. The combination is deadly, promoting explosive eruptions that produce great quantities of ash, as well as pyroclastic surges like the one that destroyed the city of Saint-Pierre in Martinique in 1902. They are also steeper than shield volcanoes, with slopes of 30–35° compared to slopes of generally 5–10°, and their loose tephra are material for dangerous lahars.

Large pieces of tephra are called volcanic bombs. Big bombs can measure more than 4 feet (1.2 meters) across and weigh several tons.

Supervolcanoes

A supervolcano is a volcano that has experienced one or more eruptions that produced over 1,000 cubic kilometers (240 cu mi) of volcanic deposits in a single explosive event.

Such eruptions occur when a very large magma chamber full of gas-rich, silicic magma is emptied in a catastrophic caldera-forming eruption. Ash flow tuffs emplaced by such eruptions are the only volcanic product with volumes rivaling those of flood basalts.

A supervolcano can produce devastation on a continental scale. Such volcanoes are able to severely cool global temperatures for many years after the eruption due to the huge volumes of sulfur and ash released into the atmosphere. They are the most dangerous type of volcano. Examples include Yellowstone Caldera in Yellowstone National Park and Valles Caldera in New Mexico (both western United States); Lake Taupo in New Zealand; Lake Toba in Sumatra, Indonesia; and Ngorongoro Crater in Tanzania. Fortunately, supervolcano eruptions are very rare events, though because of the enormous area they cover, and subsequent concealment under vegetation and glacial deposits, supervolcanoes can be difficult to identify in the geologic record without careful geologic mapping.

Submarine volcanoes

Submarine volcanoes are common features of the ocean floor. Volcanic activity during the Holocene Epoch has been documented at only 119 submarine volcanoes, but there may be more than one million geologically young submarine volcanoes on the ocean floor.

In shallow water, active volcanoes disclose their presence by blasting steam and rocky debris high above the ocean's surface. In the deep ocean basins, the tremendous weight of the water prevents the explosive release of steam and gases; however, submarine eruptions can be detected by hydrophones and by the discoloration of water because of volcanic gases. Pillow lava is a common eruptive product of submarine volcanoes and is characterized by thick sequences of discontinuous pillow-shaped masses which form under water. Even large submarine eruptions may not disturb the ocean surface, due to the rapid cooling effect and increased buoyancy in water (as compared to air), which often causes volcanic vents to form steep pillars on the ocean floor. Hydrothermal vents are common near these volcanoes, and some support peculiar ecosystems based on chemotrophs feeding on dissolved minerals. Over time, the formations created by submarine volcanoes may become so large that they break the ocean surface as new islands or floating pumice rafts.

In May and June 2018, a multitude of seismic signals were detected by earthquake monitoring agencies all over the world. They took the form of unusual humming sounds, and some of the signals detected in November of that year had a duration of up to 20 minutes. An oceanographic research campaign in May 2019 showed that the previously mysterious humming noises were caused by the formation of a submarine volcano off the coast of Mayotte.

Subglacial volcanoes

Subglacial volcanoes develop underneath icecaps. They are made up of lava plateaus capping extensive pillow lavas and palagonite. These volcanoes are also called table mountains, tuyas, or (in Iceland) mobergs.

Very good examples of this type of volcano can be seen in Iceland and in British Columbia. The origin of the term comes from Tuya Butte, which is one of the several tuyas in the area of the Tuya River and Tuya Range in northern British Columbia. Tuya Butte was the first such landform analyzed and so its name has entered the geological literature for this kind of volcanic formation.

The Tuya Mountains Provincial Park was recently established to protect this unusual landscape, which lies north of Tuya Lake and south of the Jennings River near the boundary with the Yukon Territory.

Mud volcanoes

Mud volcanoes (mud domes) are formations created by geo-excreted liquids and gases, although there are several processes which may cause such activity.

The largest structures are 10 kilometers in diameter and reach 700 meters high.

Erupted material

The material that is expelled in a volcanic eruption can be classified into three types:

Volcanic gases, a mixture made mostly of steam, carbon dioxide, and a sulfur compound (either sulfur dioxide, SO2, or hydrogen sulfide, H2S, depending on the temperature) Lava, the name of magma when it emerges and flows over the surface Tephra, particles of solid material of all shapes and sizes ejected and thrown through the air

Volcanic gases

The concentrations of different volcanic gases can vary considerably from one volcano to the next. Water vapor is typically the most abundant volcanic gas, followed by carbon dioxide and sulfur dioxide. Other principal volcanic gases include hydrogen sulfide, hydrogen chloride, and hydrogen fluoride. A large number of minor and trace gases are also found in volcanic emissions, for example hydrogen, carbon monoxide, halocarbons, organic compounds, and volatile metal chlorides.

Lava flows

Composition

The form and style of eruption of a volcano is largely determined by the composition of the lava it erupts. The viscosity (how fluid the lava is) and the amount of dissolved gas are the most important characteristics of magma, and both are largely determined by the amount of silica in the magma. Magma rich in silica is much more viscous than silica-poor magma, and silica-rich magma also tends to contain more dissolved gases.

Lava can be broadly classified into four different compositions:

If the erupted magma contains a high percentage (>63%) of silica, the lava is described as felsic. Felsic lavas (dacites or rhyolites) are highly viscous and are erupted as domes or short, stubby flows.

Lassen Peak in California is an example of a volcano formed from felsic lava and is actually a large lava dome.

Because felsic magmas are so viscous, they tend to trap volatiles (gases) that are present, which leads to explosive volcanism. Pyroclastic flows (ignimbrites) are highly hazardous products of such volcanoes, since they hug the volcano's slopes and travel far from their vents during large eruptions. Temperatures as high as 850 °C (1,560 °F) are known to occur in pyroclastic flows, which will incinerate everything flammable in their path, and thick layers of hot pyroclastic flow deposits can be laid down, often many meters thick.

Alaska's Valley of Ten Thousand Smokes, formed by the eruption of Novarupta near Katmai in 1912, is an example of a thick pyroclastic flow or ignimbrite deposit.

Volcanic ash that is light enough to be erupted high into the Earth's atmosphere as an eruption column may travel hundreds of kilometers before it falls back to ground as a fallout tuff. Volcanic gases may remain in the stratosphere for years.

Felsic magmas are formed within the crust, usually through melting of crust rock from the heat of underlying mafic magmas. The lighter felsic magma floats on the mafic magma without significant mixing.

Less commonly, felsic magmas are produced by extreme fractional crystallization of more mafic magmas.

This is a process in which mafic minerals crystallize out of the slowly cooling magma, which enriches the remaining liquid in silica.

If the erupted magma contains 52–63% silica, the lava is of intermediate composition or andesitic. Intermediate magmas are characteristic of stratovolcanoes.

They are most commonly formed at convergent boundaries between tectonic plates, by several processes. One process is hydration melting of mantle peridotite followed by fractional crystallization. Water from a subducting slab rises into the overlying mantle, lowering its melting point, particularly for the more silica-rich minerals. Fractional crystallization further enriches the magma in silica. It has also been suggested that intermediate magmas are produced by melting of sediments carried downwards by the subducted slab.

Another process is magma mixing between felsic rhyolitic and mafic basaltic magmas in an intermediate reservoir prior to emplacement or lava flow.

If the erupted magma contains <52% and >45% silica, the lava is called mafic (because it contains higher percentages of magnesium (Mg) and iron (Fe)) or basaltic. These lavas are usually hotter and much less viscous than felsic lavas. Mafic magmas are formed by partial melting of dry mantle, with limited fractional crystallization and assimilation of crustal material.

Mafic lavas occur in a wide range of settings. These include mid-ocean ridges; Shield volcanoes (such the Hawaiian Islands, including Mauna Loa and Kilauea), on both oceanic and continental crust; and as continental flood basalts.

Some erupted magmas contain =45% silica and produce ultramafic lava. Ultramafic flows, also known as komatiites, are very rare; indeed, very few have been erupted at the Earth's surface since the Proterozoic, when the planet's heat flow was higher. They are (or were) the hottest lavas, and were probably more fluid than common mafic lavas, with a viscosity less than a tenth that of hot basalt magma.

Lava texture

Mafic lava flows show two varieties of surface texture: (pronounced and pahoehoe , both Hawaiian words. A is characterized by a rough, clinkery surface and is the typical texture of cooler basalt lava flows. Pahoehoe is characterized by its smooth and often ropey or wrinkly surface and is generally formed from more fluid lava flows. Pahoehoe flows are sometimes observed to transition to ?a?a flows as they move away from the vent, but never the reverse.

More silicic lava flows take the form of block lava, where the flow is covered with angular, vesicle-poor blocks. Rhyolitic flows typically consist largely of obsidian.

Tephra

Tephra is made when magma inside the volcano is blown apart by the rapid expansion of hot volcanic gases. Magma commonly explodes as the gas dissolved in it comes out of solution as the pressure decreases when it flows to the surface. These violent explosions produce particles of material that can then fly from the volcano. Solid particles smaller than 2 mm in diameter (sand-sized or smaller) are called volcanic ash.

Tephra and other volcaniclastics (shattered volcanic material) make up more of the volume of many volcanoes than do lava flows. Volcaniclastics may have contributed as much as a third of all sedimentation in the geologic record. The production of large volumes of tephra is characteristic of explosive volcanism.

Types of volcanic eruptions

Eruption styles are broadly divided into magmatic, phreatomagmatic, and phreatic eruptions.

Magmatic eruptions

Magmatic eruptions are driven primarily by gas release due to decompression. Low-viscosity magma with little dissolved gas produces relatively gentle effusive eruptions. High-viscosity magma with a high content of dissolved gas produces violent explosive eruptions. The range of observed eruption styles is expressed from historical examples.

Hawaiian eruptions are typical of volcanoes that erupt mafic lava with a relatively low gas content. These are almost entirely effusive, producing local fire fountains and highly fluid lava flows but relatively little tephra. They are named after the Hawaiian volcanoes.

Strombolian eruptions are characterized by moderate viscosities and dissolved gas levels. They are characterized by frequent but short-lived eruptions that can produce eruptive columns hundreds of meters high. Their primary product is scoria. They are named after Stromboli.

Vulcanian eruptions are characterized by yet higher viscosities and partial crystallization of magma, which is often intermediate in composition. Eruptions take the form of short-lived explosions over the course of several hours, which destroy a central dome and eject large lava blocks and bombs. This is followed by an effusive phase that rebuilds the central dome. Vulcanian eruptions are named after Vulcano.

Peléan eruptions are more violent still, being characterized by dome growth and collapse that produces various kinds of pyroclastic flows. They are named after Mount Pelée.

Plinian eruptions are the most violent of all volcanic eruptions. They are characterized by sustained huge eruption columns whose collapse produces catastrophic pyroclastic flows. They are named after Pliny the Younger, who chronicled the Plinian eruption of Mount Vesuvius in 79 AD.

The intensity of explosive volcanism is expressed using the Volcanic Explosivity Index (VEI), which ranges from 0 for Hawaiian-type eruptions to 8 for supervolcanic eruptions.

Phreatomagmatic eruptions

Phreatomagmatic eruptions are characterized by interaction of rising magma with groundwater. They are driven by the resulting rapid buildup of pressure in the superheated groundwater.

Phreatic eruptions

Phreatic eruptions are characterized by superheating of groundwater that comes in contact with hot rock or magma. They are distinguished from phreatomagmatic eruptions because the erupted material is all country rock; no magma is erupted.

Volcanic Gases and Their Effects

Magma contains dissolved gases that are released into the atmosphere during eruptions. Gases are also released from magma that either remains below ground (for example, as an intrusion) or is rising toward the surface. In such cases, gases may escape continuously into the atmosphere from the soil, volcanic vents, fumaroles, and hydrothermal systems.

At high pressures deep beneath the earth's surface, volcanic gases are dissolved in molten rock. But as magma rises toward the surface where the pressure is lower, gases held in the melt begin to form tiny bubbles. The increasing volume taken up by gas bubbles makes the magma less dense than the surrounding rock, which may allow the magma to continue its upward journey. Closer to the surface, the bubbles increase in number and size so that the gas volume may exceed the melt volume in the magma, creating a magma foam. The rapidly expanding gas bubbles of the foam can lead to explosive eruptions in which the melt is fragmented into pieces of volcanic rock, known as tephra. If the molten rock is not fragmented by explosive activity, a lava flow will be generated.

Together with the tephra and entrained air, volcanic gases can rise tens of kilometers into Earth's atmosphere during large explosive eruptions. Once airborne, the prevailing winds may blow the eruption cloud hundreds to thousands of kilometers from a volcano. The gases spread from an erupting vent primarily as acid aerosols (tiny acid droplets), compounds attached to tephra particles, and microscopic salt particles.

Volcanic gases undergo a tremendous increase in volume when magma rises to the Earth's surface and erupts. For example, consider what happens if one cubic meter of 900°C rhyolite magma containing five percent by weight of dissolved water were suddenly brought from depth to the surface. The one cubic meter of magma now would occupy a volume of 670 m3 as a mixture of water vapor and magma at atmospheric pressure (Sparks et. al., 1997)! The one meter cube at depth would increase to 8.75 m on each side at the surface. Such enormous expansion of volcanic gases, primarily water, is the main driving force of explosive eruptions.

Sulfur Dioxide (SO2)

The effects of SO2 on people and the environment vary widely depending on (1) the amount of gas a volcano emits into the atmosphere; (2) whether the gas is injected into the troposphere or stratosphere; and (3) the regional or global wind and weather pattern that disperses the gas. Sulfur dioxide (SO2) is a colorless gas with a pungent odor that irritates skin and the tissues and mucous membranes of the eyes, nose, and throat. Sulfur dioxide chiefly affects upper respiratory tract and bronchi. The World Health Organization recommends a concentration of no greater than 0.5 ppm over 24 hours for maximum exposure. A concentration of 6-12 ppm can cause immediate irritation of the nose and throat; 20 ppm can cause eye irritation; 10,000 ppm will irritate moist skin within minutes.

Emission rates of SO2 from an active volcano range from <20 tonnes/day to >10 million tonnes/day according to the style of volcanic activity and type and volume of magma involved. For example, the large explosive eruption of Mount Pinatubo on 15 June 1991 expelled 3-5 km3 of dacite magma and injected about 20 million metric tons of SO2 into the stratosphere. The sulfur aerosols resulted in a 0.5-0.6°C cooling of the Earth's surface in the Northern Hemisphere. The sulfate aerosols also accelerated chemical reactions that, together with the increased stratospheric chlorine levels from human-made chlorofluorocarbon (CFC) pollution, destroyed ozone and led to some of the lowest ozone levels ever observed in the atmosphere.

At Kilauea Volcano, the recent effusive eruption of about 0.0005 km3/day (500,000 m3) of basalt magma releases about 2,000 tonnes of SO2 into the lower troposphere.

Downwind from the vent, acid rain and air pollution is a persistent health problem when the volcano is erupting.

SO2 Causes air Pollution Volcanic Smog

Eruptions of Kilauea Volcano release large quantities of sulfur dioxide gas into the atmosphere that can lead to volcanic air pollution on the Island of Hawai'i. Sulfur dioxide gas reacts chemically with sunlight, oxygen, dust particles, and water to form volcanic smog known as vog.

SO2 effects Earth's Surface Temperature Global Cooling and Ozone Depletion

Measurements from recent eruptions such as Mount St. Helens, Washington (1980), El Chichon, Mexico (1982), and Mount Pinatubo, Philippines (1991), clearly show the importance of sulfur aerosols in modifying climate, warming the stratosphere, and cooling the troposphere. Research has also shown that the liquid drops of sulfuric acid promote the destruction of the Earth's ozone layer.

Hydrogen Sulfide (H2S)

Hydrogen sulfide (H2S) is a colorless, flammable gas with a strong offensive odor. It is sometimes referred to as sewer gas. At low concentrations it can irritate the eyes and acts as a depressant; at high concentrations it can cause irritation of the upper respiratory tract and, during long exposure, pulmonary edema. A 30-minute exposure to 500 ppm results in headache, dizziness, excitement, staggering gait, and diarrhea, followed sometimes by bronchitis or bronchopneumonia.

Carbon Dioxide (CO2)

Volcanoes release more than 130 million tonnes of CO2 into the atmosphere every year. This colorless, odorless gas usually does not pose a direct hazard to life because it typically becomes diluted to low concentrations very quickly whether it is released continuously from the ground or during episodic eruptions. But in certain circumstances, CO2 may become concentrated at levels lethal to people and animals. Carbon dioxide gas is heavier than air and the gas can flow into in low-lying areas; breathing air with more than 30% CO2 can quickly induce unconsciousness and cause death. In volcanic or other areas where CO2 emissions occur, it is important to avoid small depressions and low areas that might be CO2 traps. The boundary between air and lethal gas can be extremely sharp; even a single step upslope may be adequate to escape death.

When a burning piece of cloth is lowered into a hole that has a high concentration of CO2, the fire goes out. Such a condition can be lethal to people and animals.

Air with 5% CO2 causes perceptible increased respiration; 6-10% results in shortness of breath, headaches, dizziness, sweating, and general restlessness; 10-15% causes impaired coordination and abrupt muscle contractions; 20-30% causes loss of consciousness and convulsions; over 30% can cause death (Hathaway et. al., 1991).

Hydrogen Chloride (HCl)

Chlorine gas is emitted from volcanoes in the form of hydrochloric acid (HCl). Exposure to the gas irritates mucous membranes of the eyes and respiratory tract. Concentrations over 35 ppm cause irritation of the throat after short exposure; >100 ppm results in pulmonary edema, and often laryngeal spasm. It also causes acid rain downwind from volcanoes because HCl is extremely soluble in condensing water droplets and it is a very "strong acid" (it dissociates extensively to give H+ ions in the droplets).

Hydrogen Fluoride (HF)

Fluorine is a pale yellow gas that attaches to fine ash particles, coats grass, and pollutes streams and lakes. Exposure to this powerful caustic irritant can cause conjunctivitis, skin irritation, bone degeneration and mottling of teeth. Excess fluorine results in a significant cause of death and injury in livestock during ash eruptions. Even in areas that receive just a millimeter of ash, poisoning can occur where the fluorine content of dried grass exceeds 250 ppm. Animals that eat grass coated with fluorine-tainted ash are poisoned. Small amounts of fluorine can be beneficial, but excess fluorine causes fluorosis, an affliction that eventually kills animals by destroying their bones. It also promotes acid rain effects downwind of volcanoes, like HCl.

Secondary Gas Emissions

Another type of gas release occurs when lava flows reach the ocean. Extreme heat from molten lava boils and vaporizes seawater, leading to a series of chemical reactions. The boiling and reactions produce a large white plume, locally known as lava haze or laze, containing a mixture of hydrochloric acid and concentrated seawater.

Laze Plumes are Very Acidic

Extreme heat from lava entering the sea rapidly boils and vaporizes seawater, leading to a series of chemical reactions. The boiling and reactions produce a large white plume, locally known as lava haze or laze, which contains a mixture of hydrochloric acid (HCl) and concentrated seawater. This is a short-lived local phenomenon that only affects people or vegetation directly under the plume.

The hydrochloric acid (HCl) comes from the breakdown of seawater-derived chlorides during sudden boiling. Because the lava is largely degassed by the time it reaches the sea, any HCL coming from it is insignificant by comparison. Analyzed samples of the plume show that is is a brine with a salinity of about 2.3 times that of seawater and a pH of 1.5-2.0.

Volcanic Features

The most common perception of a volcano is of a conical mountain, spewing lava and poisonous gases from a crater at its summit. This describes just one of many types of volcano, and the features of volcanoes are much more complicated. The structure and behavior of volcanoes depends on a number of factors. Some volcanoes have rugged peaks formed by lava domes rather than a summit crater, whereas others present landscape features such as massive plateaus. Vents that issue volcanic material (lava, which is what magma is called once it has escaped to the surface, and ash) and gases (mainly steam and magmatic gases) can be located anywhere on the landform.

Many of these vents give rise to smaller cones such as flank of Hawaii's Kîlauea. Other types of volcano include cryovolcanoes (or ice volcanoes), particularly on some moons of Jupiter, Saturn and Neptune; and mud volcanoes, which are formations often not associated with known magmatic activity. Active mud volcanoes tend to involve temperatures much lower than those of igneous volcanoes, except when a mud volcano is actually a vent of an igneous volcano.

Fissure Vents

Volcanic fissure vents are flat, linear cracks through which lava emerges.

Shield Volcanoes

Shield volcanoes, so named for their broad, shield-like profiles, are formed by the eruption of low-viscosity lava that can flow a great distance from a vent, but not generally explode catastrophically. Since low-viscosity magma is typically low in silica, shield volcanoes are more common in oceanic than continental settings. The Hawaiian volcanic chain is a series of shield cones, and they are common in Iceland, as well.

Lava Domes

Lava domes are built by slow eruptions of highly viscous lavas. They are sometimes formed within the crater of a previous volcanic eruption (as in Mount Saint Helens), but can also form independently, as in the case of Lassen Peak. Like stratovolcanoes, they can produce violent, explosive eruptions, but their lavas generally do not flow far from the originating vent.

Cryptodomes

Cryptodomes are formed when viscous lava forces its way up and causes a bulge. The 1980 eruption of Mount St. Helens was an example. Lava was under great pressure and forced a bulge in the mountain, which was unstable and slid down the north side.

Volcanic Cones (cinder cones)

Volcanic cones or cinder cones are the result from eruptions that erupt mostly small pieces of scoria and pyroclastics (both resemble cinders, hence the name of this volcano type) that build up around the vent. These can be relatively short-lived eruptions that produce a cone-shaped hill perhaps 30 to 400 meters high. Most cinder cones erupt only once. Cinder cones may form as flank vents on larger volcanoes, or occur on their own. Parícutin in Mexico and Sunset Crater in Arizona are examples of cinder cones. In New Mexico, Caja del Rio is a volcanic field of over 60 cinder cones.

Stratovolcanoes or composite volcanoes are tall conical mountains composed of lava flows and other ejecta in alternate layers, the strata that give rise to the name. Stratovolcanoes are also known as composite volcanoes, created from several structures during different kinds of eruptions. Strato/composite volcanoes are made of cinders, ash and lava. Cinders and ash pile on top of each other, lava flows on top of the ash, where it cools and hardens, and then the process begins again. Classic examples include Mt. Fuji in Japan, Mayon Volcano in the Philippines, and Mount Vesuvius and Stromboli in Italy.

In recorded history, explosive eruptions by stratovolcanoes have posed the greatest hazard to civilizations, as ash is produced by an explosive eruption. No supervolcano erupted in recorded history. Shield volcanoes have not an enormous pressure build up from the lava flow. Fissure vents and monogenetic volcanic fields (volcanic cones) have not powerful explosive eruptions, as they are many times under extension. Stratovolcanoes (30–35°) are steeper than shield volcanoes (generally 5–10°), their loose tephraare material for dangerous lahars.

Supervolcanoes

A supervolcano is a large volcano that usually has a large caldera and can potentially produce devastation on an enormous, sometimes continental, scale. Such eruptions would be able to cause severe cooling of global temperatures for many years afterwards because of the huge volumes of sulfur and ash erupted. They are the most dangerous type of volcano. Examples include Yellowstone Caldera in Yellowstone National Park and Valles Caldera in New Mexico (both western United States), Lake Taupo in New Zealand, Lake Toba in Sumatra, Indonesia and Ngorogoro Crater in Tanzania, Krakatoa near Java and Sumatra,Indonesia. Supervolcanoes are hard to identify centuries later, given the enormous areas they cover. Large igneous provinces are also considered supervolcanoes because of the vast amount of basalt lava erupted, but are non-explosive.

Submarine Volcanoes

Submarine volcanoes are common features on the ocean floor. Some are active and, in shallow water, disclose their presence by blasting steam and rocky debris high above the surface of the sea. Many others lie at such great depths that the tremendous weight of the water above them prevents the explosive release of steam and gases, although they can be detected by hydrophones and discoloration of water because of volcanic gases. Pumice rafts may also appear.

Even large submarine eruptions may not disturb the ocean surface. Because of the rapid cooling effect of water as compared to air, and increased buoyancy, submarine volcanoes often form rather steep pillars over their volcanic vents as compared to above-surface volcanoes. They may become so large that they break the ocean surface as new islands. Pillow lava is a common eruptive product of submarine volcanoes. Hydrothermal vents are common near these volcanoes, and some support peculiar ecosystems based on dissolved minerals.

Subglacial Volcanoes

Subglacial volcanoes develop underneath icecaps. They are made up of flat lava which flows at the top of extensive pillow lavas and palagonite. When the icecap melts, the lavas on the top collapse, leaving a flat-topped mountain. These volcanoes are also called table mountains, tuyas or (uncommonly) mobergs. Very good examples of this type of volcano can be seen in Iceland, however, there are also tuyas in British Columbia.

The origin of the term comes from Tuya Butte, which is one of the several tuyas in the area of the Tuya River and Tuya Range in northern British Columbia. Tuya Butte was the first such landform analyzed and so its name has entered the geological literature for this kind of volcanic formation. The Tuya Mountains Provincial Park was recently established to protect this unusual landscape, which lies north of Tuya Lake and south of the Jennings River near the boundary with the Yukon Territory.

Mud Volcanoes

Mud volcanoes or mud domes are formations created by geo-excreted liquids and gases, although there are several processes which may cause such activity. The largest structures are 10 kilometers in diameter and reach 700 meters high.

Lava Texture

Two types of lava are named according to the surface texture: pronounce and pâhoehoe, both Hawaiian words.It is characterized by a rough, clinkery surface and is the typical texture of viscous lava flows. However, even basaltic or mafic flows can be erupted as »a»a flows, particularly if the eruption rate is high and the slope is steep.

Pâhoehoe is characterized by its smooth and often ropey or wrinkly surface and is generally formed from more fluid lava flows. Usually, only mafic flows will erupt as pâhoehoe, since they often erupt at higher temperatures or have the proper chemical make-up to allow them to flow with greater fluidity.

Volcanic Activity

A popular way of classifying magmatic volcanoes is by their frequency of eruption, with those that erupt regularly called active, those that have erupted in historical times but are now quiet called dormant, and those that have not erupted in historical times called extinct. However, these popular classifications—extinct in particular—are practically meaningless to scientists. They use classifications which refer to a particular volcano's formative and eruptive processes and resulting shapes, which was explained above.

There is no real consensus among volcanologists on how to define an "active" volcano. The lifespan of a volcano can vary from months to several million years, making such a distinction sometimes meaningless when compared to the lifespans of humans or even civilizations. For example, many of Earth's volcanoes have erupted dozens of times in the past few thousand years but are not currently showing signs of eruption. Given the long lifespan of such volcanoes, they are very active. By human lifespans, however, they are not.

Scientists usually consider a volcano to be erupting or likely to erupt if it is currently erupting, or showing signs of unrest such as unusual earthquake activity or significant new gas emissions. Most scientists consider a volcano active if it has erupted in holocene times. Historic times is another timeframe for active. But it is important to note that the span of recorded history differs from region to region. In China and the Mediterranean, recorded history reaches back more than 3,000 years but in the Pacific Northwest of the United States and Canada, it reaches back less than 300 years, and in Hawaii and New Zealand, only around 200 years. The Smithsonian Global Volcanism Program's definition of active is having erupted within the last 10,000 years (the 'holocene' period).

Presently there are about 500 active volcanoes in the world – the majority following along the Pacific 'Ring of Fire' – and around 50 of these erupt each year. The United States is home to 50 active volcanoes. There are more than 1,500 potentially active volcanoes. An estimated 500 million people live near active volcanoes.

Extinct

Extinct volcanoes are those that scientists consider unlikely to erupt again, because the volcano no longer has a lava supply. Examples of extinct volcanoes are many volcanoes on the Hawaiian – Emperor seamount chain in the Pacific Ocean, Hohentwiel, Shiprock and the Zuidwal volcano in the Netherlands. Edinburgh Castle in Scotland is famously located atop an extinct volcano. Otherwise, whether a volcano is truly extinct is often difficult to determine. Since "supervolcano" calderas can have eruptive lifespans sometimes measured in millions of years, a caldera that has not produced an eruption in tens of thousands of years is likely to be considered dormant instead of extinct.

Dormant

It is difficult to distinguish an extinct volcano from a dormant one. Volcanoes are often considered to be extinct if there are no written records of its activity. Nevertheless, volcanoes may remain dormant for a long period of time. For example, Yellowstone has a repose/recharge period of around 700 ka, and Toba of around 380 ka. Vesuvius was described by Roman writers as having been covered with gardens and vineyards before its famous eruption of AD 79, which destroyed the towns of Herculaneum and Pompeii. Before its catastrophic eruption of 1991, Pinatubo was an inconspicuous volcano, unknown to most people in the surrounding areas. Two other examples are the long-dormant Soufrière Hills volcano on the island of Montserrat, thought to be extinct before activity resumed in 1995 and Fourpeaked Mountain in Alaska, which, before its September 2006 eruption, had not erupted since before 8000 BC and had long been thought to be extinct.

Technical classification of volcanoes

Volcanic-alert Level

The three common popular classifications of volcanoes can be subjective and some volcanoes thought to have been extinct have announced to the world they were just pretending. To help prevent citizens from falsely believing they are not at risk when living on or near a volcano, countries have adopted new classifications to describe the various levels and stages of volcanic activity. Some alert systems use different numbers or colors to designate the different stages. Other systems use colors and words. Some systems use a combination of both.

Volcano Warning Schemes of the United States

The United States Geological Survey (USGS) has adopted a common system nationwide for characterizing the level of unrest and eruptive activity at volcanoes. The new volcano alert-level system classifies volcanoes now as being in a normal, advisory, watch or warning stage. Additionally, colors are used to denote the amount of ash produced. Details of the US system can be found at Volcano warning schemes of the United States.

The Decade Volcanoes are 16 volcanoes identified by the International Association of Volcanology and Chemistry of the Earth's Interior(IAVCEI) as being worthy of particular study in light of their history of large, destructive eruptions and proximity to populated areas. They are named Decade Volcanoes because the project was initiated as part of the United Nations-sponsored International Decade for Natural Disaster Reduction. The 16 current Decade Volcanoes are :

- Avachinsky-Koryaksky, Kamchatka, Russia
- Nevado de Colima, Jalisco and Colima, Mexico
- Mount Etna, Sicily, Italy
- Galeras, Nariño, Colombia
- Mauna Loa, Hawaii, USA
- Mount Merapi, Central Java, Indonesia
- Mount Nyiragongo, Democratic Republic of the Congo
- Mount Rainier, Washington, USA
- Sakurajima, Kagoshima Prefecture, Japan
- Santa Maria/Santiaguito, Guatemala
- Santorini, Cyclades, Greece
- Taal Volcano, Luzon, Philippines
- Teide, Canary Islands, Spain
- Ulawun, New Britain, Papua New Guinea
- Mount Unzen, Nagasaki Prefecture, Japan
- Vesuvius, Naples, Italy

Effects of Volcanoes

There are many different types of volcanic eruptions and associated activity: phreatic eruptions(steam-generated eruptions), explosive eruption of high-silica lava (e.g., rhyolite), effusive eruption of low-silica lava (e.g., basalt), pyroclastic flows, lahars (debris flow) and carbon dioxide emission. All of these activities can pose a hazard to humans. Earthquakes, hot springs, fumaroles, mud potsand geysers often accompany volcanic activity.

The concentrations of different volcanic gases can vary considerably from one volcano to the next.Water vapor is typically the most abundant volcanic gas, followed by carbon dioxide and sulfur dioxide. Other principal volcanic gases include hydrogen sulfide, hydrogen chloride, and hydrogen fluoride. A large number of minor and trace gases are also found in volcanic emissions, for examplehydrogen, carbon monoxide, halocarbons, organic compounds, and volatile metal chlorides.

Large, explosive volcanic eruptions inject water vapor (H2O), carbon dioxide (CO2), sulfur dioxide (SO2), hydrogen chloride (HCl), hydrogen fluoride (HF) and ash (pulverized rock and pumice) into thestratosphere to heights of 16–32 kilometres (10–20 mi) above the Earth's surface. The most significant impacts from these injections come from the conversion of sulfur dioxide to sulfuric acid (H2SO4), which condenses rapidly in the stratosphere to form fine sulfate aerosols. The aerosols increase the Earth's albedo—its reflection of radiation from the Sun back into space – and thus cool the Earth's lower atmosphere or troposphere; however, they also absorb heat radiated up from the Earth, thereby warming the stratosphere. Several eruptions during the past century have caused a decline in the average temperature at the Earth's surface of up to half a degree (Fahrenheit scale) for periods of one to three years — sulfur dioxide from the eruption of Huaynaputina probably caused the Russian famine of 1601–1603.

One proposed volcanic winter happened c. 70,000 years ago following the supereruption of Lake Toba on Sumatra island in Indonesia. According to the Toba catastrophe theory to which some anthropologists and archeologists subscribe, it had global consequences, killing most humans then alive and creating a population bottleneck that affected the genetic inheritance of all humans today. The 1815 eruption ofMount Tambora created global climate anomalies that became known as the "Year Without a Summer" because of the effect on North American and European weather. Agricultural crops failed and livestock died in much of the Northern Hemisphere, resulting in one of the worst famines of the 19th century. The freezing winter of 1740–41, which led to widespread famine in northern Europe, may also owe its origins to a volcanic eruption.

It has been suggested that volcanic activity caused or contributed to the End-Ordovician, Permian-Triassic, Late Devonian mass extinctions, and possibly others. The massive eruptive event which formed the Siberian Traps, one of the largest known volcanic events of the last 500 million years of Earth's geological history, continued for a million years and is considered to be the likely cause of the "Great Dying" about 250 million years ago, which is estimated to have killed 90% of species existing at the time.

The sulfate aerosols also promote complex chemical reactions on their surfaces that alter chlorine and nitrogen chemical species in the stratosphere. This effect, together with increased stratospheric chlorine levels from chlorofluorocarbon pollution, generates chlorine monoxide (ClO), which destroys ozone (O3). As the aerosols grow and coagulate, they settle down into the upper troposphere where they serve as nuclei for cirrus clouds and further modify the Earth's radiation balance. Most of the hydrogen chloride (HCl) and hydrogen fluoride (HF) are dissolved in water droplets in the eruption cloud and quickly fall to the ground as acid rain. The injected ash also falls rapidly from the stratosphere; most of it is removed within several days to a few weeks. Finally, explosive volcanic eruptions release the greenhouse gas carbon dioxide and thus provide a deep source of carbon for biogeochemical cycles.

Gas emissions from volcanoes are a natural contributor to acid rain. Volcanic activity releases about 130 to 230 teragrams (145 million to 255 million short tons) of carbon dioxide each year. Volcanic eruptions may inject aerosols into the Earth's atmosphere. Large injections may cause visual effects such as unusually colorful sunsets and affect global climatemainly by cooling it. Volcanic eruptions also provide the benefit of adding nutrients to soil through the weathering process of volcanic rocks. These fertile soils assist the growth of plants and various crops. Volcanic eruptions can also create new islands, as the magma cools and solidifies upon contact with the water.

Ash thrown into the air by eruptions can present a hazard to aircraft, especially jet aircraft where the particles can be melted by the high operating temperature. Dangerous encounters in 1982 after the eruption of Galunggung in Indonesia, and 1989 after the eruption of Mount Redoubt in Alaska raised awareness of this phenomenon. Nine Volcanic Ash Advisory Centerswere established by the International Civil Aviation Organization to monitor ash clouds and advise pilots accordingly. The 2010 eruptions of Eyjafjallajökull caused major disruptions to air travel in Europe.

The Earth's Moon has no large volcanoes and no current volcanic activity, although recent evidence suggests it may still possess a partially molten core. However, the Moon does have many volcanic features such as maria (the darker patches seen on the moon), rilles and domes.

The planet Venus has a surface that is 90% basalt, indicating that volcanism played a major role in shaping its surface. The planet may have had a major global resurfacing event about 500 million years ago, from what scientists can tell from the density of impact craters on the surface. Lava flows are widespread and forms of volcanism not present on Earth occur as well. Changes in the planet's atmosphere and observations of lightning have been attributed to ongoing volcanic eruptions, although there is no confirmation of whether or not Venus is still volcanically active. However, radar sounding by the Magellan probe revealed evidence for comparatively recent volcanic activity at Venus's highest volcano Maat Mons, in the form of ash flows near the summit and on the northern flank.

There are several extinct volcanoes on Mars, four of which are vast shield volcanoes far bigger than any on Earth. They include Arsia Mons, Ascraeus Mons,

Hecates Tholus, Olympus Mons, and Pavonis Mons. These volcanoes have been extinct for many millions of years, but the European Mars Express spacecraft has found evidence that volcanic activity may have occurred on Mars in the recent past as well.

Jupiter's moon Io is the most volcanically active object in the solar system because of tidal interaction with Jupiter. It is covered with volcanoes that erupt sulfur, sulfur dioxide and silicate rock, and as a result, Io is constantly being resurfaced. Its lavas are the hottest known anywhere in the solar system, with temperatures exceeding 1,800 K (1,500 °C). In February 2001, the largest recorded volcanic eruptions in the solar system occurred on Io. Europa, the smallest of Jupiter's Galilean moons, also appears to have an active volcanic system, except that its volcanic activity is entirely in the form of water, which freezes into ice on the frigid surface. This process is known as cryovolcanism, and is apparently most common on the moons of the outer planets of the solar system.

In 1989 the Voyager 2 spacecraft observed cryovolcanoes (ice volcanoes) on Triton, a moon of Neptune, and in 2005 the Cassini–Huygens probe photographed fountains of frozen particles erupting from Enceladus, a moon of Saturn. The ejecta may be composed of water, liquid nitrogen, dust, or methane compounds. Cassini–Huygens also found evidence of a methane-spewing cryovolcano on the Saturnian moon Titan, which is believed to be a significant source of the methane found in its atmosphere. It is theorized that cryovolcanism may also be present on the Kuiper Belt Object Quaoar.

A 2010 study of the exoplanet COROT-7b, which was detected by transit in 2009, studied that tidal heating from the host star very close to the planet and neighboring planets could generate intense volcanic activity similar to Io.

Earthquake

An earthquake (also known as a quake, tremor or temblor) is the shaking of the surface of the Earth resulting from a sudden release of energy in the Earth's lithosphere that creates seismic waves. Earthquakes can range in intensity, from those that are so weak that they cannot be felt, to those violent enough to propel objects and people into the air and wreak destruction across entire cities. The seismicity, or seismic activity, of an area is the frequency, type, and size of earthquakes experienced over a particular time period. The word tremor is also used for non-earthquake seismic rumbling.

At the Earth's surface, earthquakes manifest themselves by shaking and displacing or disrupting the ground. When the epicenter of a large earthquake is located offshore, the seabed may be displaced sufficiently to cause a tsunami. Earthquakes can also trigger landslides and, occasionally, volcanic activity.

In its most general sense, the word earthquake is used to describe any seismic event—whether natural or caused by humans—that generates seismic waves. Earthquakes are caused mostly by rupture of geological faults but also by other events such as volcanic activity, landslides, mine blasts, and nuclear tests. An earthquake's point of initial rupture is called its hypocenter or focus. The epicenter is the point at ground level directly above the hypocenter.

Naturally occurring earthquakes

Tectonic earthquakes occur anywhere in the earth where there is sufficient stored elastic strain energy to drive fracture propagation along a fault plane. The sides of a fault move past each other smoothly and aseismically only if there are no irregularities or asperities along the fault surface that increase the frictional resistance. Most fault surfaces do have such asperities, which leads to a form of stick-slip behavior. Once the fault has locked, continued relative motion between the plates leads to increasing stress and, therefore, stored strain energy in the volume around the fault surface. This continues until the stress has risen sufficiently to break through the asperity, suddenly allowing sliding over the locked portion of the fault, releasing the stored energy.

This energy is released as a combination of radiated elastic strain seismic waves, frictional heating of the fault surface, and cracking of the rock, thus causing an earthquake. This process of gradual build-up of strain and stress punctuated by occasional sudden earthquake failure is referred to as the elastic-rebound theory. It is estimated that only 10 percent or less of an earthquake's total energy is radiated as seismic energy. Most of the earthquake's energy is used to power the earthquake fracture growth or is converted into heat generated by friction. Therefore, earthquakes lower the Earth's available elastic potential energy and raise its temperature, though these changes are negligible compared to the conductive and convective flow of heat out from the Earth's deep interior.

Earthquake fault types

There are three main types of fault, all of which may cause an interplate earthquake: normal, reverse (thrust), and strike-slip. Normal and reverse faulting are examples of dip-slip, where the displacement along the fault is in the direction of dip and where movement on them involves a vertical component. Normal faults occur mainly in areas where the crust is being extended such as a divergent boundary. Reverse faults occur in areas where the crust is being shortened such as at a convergent boundary. Strike-slip faults are steep structures where the two sides of the fault slip horizontally past each other; transform boundaries are a particular type of strike-slip fault. Many earthquakes are caused by movement on faults that have components of both dip-slip and strike-slip; this is known as oblique slip.

Reverse faults, particularly those along convergent plate boundaries, are associated with the most powerful earthquakes, megathrust earthquakes, including almost all of those of magnitude 8 or more. Megathrust earthquakes are responsible for about 90% of the total seismic moment released worldwide.

Strike-slip faults, particularly continental transforms, can produce major earthquakes up to about magnitude 8. Earthquakes associated with normal faults are generally less than magnitude 7. For every unit increase in magnitude, there is a roughly thirtyfold increase in the energy released. For instance, an earthquake of magnitude 6.0 releases approximately 32 times more energy than a 5.0 magnitude earthquake and a 7.0 magnitude earthquake releases 1,000 times more energy than a 5.0 magnitude earthquake. An 8.6 magnitude earthquake releases the same amount of energy as 10,000 atomic bombs of the size used in World War II.

This is so because the energy released in an earthquake, and thus its magnitude, is proportional to the area of the fault that ruptures and the stress drop. Therefore, the longer the length and the wider the width of the faulted area, the larger the resulting magnitude. The topmost, brittle part of the Earth's crust, and the cool slabs of the tectonic plates that are descending into the hot mantle, are the only parts of our planet that can store elastic energy and release it in fault ruptures. Rocks hotter than about 300 °C (572 °F) flow in response to stress; they do not rupture in earthquakes.

The maximum observed lengths of ruptures and mapped faults (which may break in a single rupture) are approximately 1,000 km (620 mi). Examples are the earthquakes in Alaska (1957), Chile (1960), and Sumatra (2004), all in subduction zones. The longest earthquake ruptures on strike-slip faults, like the San Andreas Fault (1857, 1906), the North Anatolian Fault in Turkey (1939), and the Denali Fault in Alaska (2002), are about half to one third as long as the lengths along subducting plate margins, and those along normal faults are even shorter.

The most important parameter controlling the maximum earthquake magnitude on a fault, however, is not the maximum available length, but the available width because the latter varies by a factor of 20. Along converging plate margins, the dip angle of the rupture plane is very shallow, typically about 10 degrees.

Thus, the width of the plane within the top brittle crust of the Earth can become 50–100 km (31–62 mi) (Japan, 2011; Alaska, 1964), making the most powerful earthquakes possible.

Strike-slip faults tend to be oriented near vertically, resulting in an approximate width of 10 km (6.2 mi) within the brittle crust.

Thus, earthquakes with magnitudes much larger than 8 are not possible. Maximum magnitudes along many normal faults are even more limited because many of them are located along spreading centers, as in Iceland, where the thickness of the brittle layer is only about six kilometres (3.7 mi).

In addition, there exists a hierarchy of stress levels in the three fault types. Thrust faults are generated by the highest, strike-slip by intermediate, and normal faults by the lowest stress levels.

This can easily be understood by considering the direction of the greatest principal stress, the direction of the force that "pushes" the rock mass during the faulting. In the case of normal faults, the rock mass is pushed down in a vertical direction, thus the pushing force (greatest principal stress) equals the weight of the rock mass itself. In the case of thrusting, the rock mass "escapes" in the direction of the least principal stress, namely upward, lifting the rock mass, and thus, the overburden equals the least principal stress. Strike-slip faulting is intermediate between the other two types described above. This difference in stress regime in the three faulting environments can contribute to differences in stress drop during faulting, which contributes to differences in the radiated energy, regardless of fault dimensions.

Earthquakes away from plate boundaries

Where plate boundaries occur within the continental lithosphere, deformation is spread out over a much larger area than the plate boundary itself. In the case of the

San Andreas fault continental transform, many earthquakes occur away from the plate boundary and are related to strains developed within the broader zone of deformation caused by major irregularities in the fault trace (e.g., the "Big bend" region). The Northridge earthquake was associated with movement on a blind thrust within such a zone. Another example is the strongly oblique convergent plate boundary between the Arabian and Eurasian plates where it runs through the northwestern part of the Zagros Mountains. The deformation associated with this plate boundary is partitioned into nearly pure thrust sense movements perpendicular to the boundary over a wide zone to the southwest and nearly pure strike-slip motion along the Main Recent Fault close to the actual plate boundary itself. This is demonstrated by earthquake focal mechanisms.

All tectonic plates have internal stress fields caused by their interactions with neighboring plates and sedimentary loading or unloading (e.g., deglaciation).

These stresses may be sufficient to cause failure along existing fault planes, giving rise to intraplate earthquakes.

Shallow-focus and deep-focus earthquakes

The majority of tectonic earthquakes originate in the ring of fire at depths not exceeding tens of kilometers. Earthquakes occurring at a depth of less than 70 km (43 mi) are classified as "shallow-focus" earthquakes, while those with a focal-depth between 70 and 300 km (43 and 186 mi) are commonly termed "mid-focus" or "intermediate-depth" earthquakes. In Subduction zones, where older and colder oceanic crust descends beneath another tectonic plate, deep-focus earthquakes may occur at much greater depths (ranging from 300 to 700 km (190 to 430 mi)).

These seismically active areas of subduction are known as Wadati–Benioff zones. Deep-focus earthquakes occur at a depth where the subducted lithosphere should no longer be brittle, due to the high temperature and pressure. A possible mechanism for the generation of deep-focus earthquakes is faulting caused by olivine undergoing a phase transition into a spinel structure.

Earthquakes and volcanic activity

Earthquakes often occur in volcanic regions and are caused there, both by tectonic faults and the movement of magma in volcanoes. Such earthquakes can serve as an early warning of volcanic eruptions, as during the 1980 eruption of Mount St. Helens.

Earthquake swarms can serve as markers for the location of the flowing magma throughout the volcanoes. These swarms can be recorded by seismometers and tiltmeters (a device that measures ground slope) and used as sensors to predict imminent or upcoming eruptions.

Rupture dynamics

A tectonic earthquake begins by an initial rupture at a point on the fault surface, a process known as nucleation. The scale of the nucleation zone is uncertain, with some evidence, such as the rupture dimensions of the smallest earthquakes, suggesting that it is smaller than 100 m (330 ft) while other evidence, such as a slow component revealed by low-frequency spectra of some earthquakes, suggest that it is larger. The possibility that the nucleation involves some sort of preparation process is supported by the observation that about 40% of earthquakes are preceded by foreshocks. Once the rupture has initiated, it begins to propagate along the fault surface. The mechanics of this process are poorly understood, partly because it is difficult to recreate the high sliding velocities in a laboratory. Also, the effects of strong ground motion make it very difficult to record information close to a nucleation zone.

Rupture propagation is generally modeled using a fracture mechanics approach, likening the rupture to a propagating mixed mode shear crack. The rupture velocity is a function of the fracture energy in the volume around the crack tip, increasing with decreasing fracture energy. The velocity of rupture propagation is orders of magnitude faster than the displacement velocity across the fault. Earthquake ruptures typically propagate at velocities that are in the range 70–90% of the S-wave velocity, which is independent of earthquake size. A small subset of earthquake ruptures appear to have propagated at speeds greater than the S-wave velocity. These supershear earthquakes have all been observed during large strike-slip events. The unusually wide zone of coseismic damage caused by the 2001 Kunlun earthquake has been attributed to the effects of the sonic boom developed in such earthquakes. Some earthquake ruptures travel at unusually low velocities and are referred to as slow earthquakes. A particularly dangerous form of slow earthquake is the tsunami earthquake, observed where the relatively low felt intensities, caused by the slow propagation speed of some great earthquakes, fail to alert the population of the neighboring coast, as in the 1896 Sanriku earthquake.

Co-seismic overpressuring and effect of pore pressure

During an earthquake, high temperatures can develop at the fault plane so increasing pore pressure consequently to vaporization of the ground water already contained within rock.

In the coseismic phase, such increase can significantly affect slip evolution and speed and, furthermore, in the post-seismic phase it can control the Aftershock sequence because, after the main event, pore pressure increase slowly propagates into the surrounding fracture network.

From the point of view of the Mohr-Coulomb strength theory, an increase in fluid pressure reduces the normal stress acting on the fault plane that holds it in place, and fluids can exert a lubricating effect. As thermal overpressurization may provide positive feedback between slip and strength fall at the fault plane, a common opinion is that it may enhance the faulting process instability. After the mainshock, the pressure gradient between the fault plane and the neighboring rock causes a fluid flow which increases pore pressure in the surrounding fracture networks; such increase may trigger new faulting processes by reactivating adjacent faults, giving rise to aftershocks.

Analogously, artificial pore pressure increase, by fluid injection in Earth's crust, may induce seismicity.

Tidal forces

Tides may induce some seismicity.

Earthquake clusters

Most earthquakes form part of a sequence, related to each other in terms of location and time.

Most earthquake clusters consist of small tremors that cause little to no damage, but there is a theory that earthquakes can recur in a regular pattern.

Aftershocks

Magnitude of the Central Italy earthquakes of August and October 2016 and January 2017 and the aftershocks (which continued to occur after the period shown here) An aftershock is an earthquake that occurs after a previous earthquake, the mainshock. Rapid changes of stress between rocks, and the stress from the original earthquake are the main causes of these aftershocks, along with the crust around the ruptured fault plane as it adjusts to the effects of the mainshock.

An aftershock is in the same region of the main shock but always of a smaller magnitude, however they can still be powerful enough to cause even more damage to buildings that were already previously damaged from the original quake.

If an aftershock is larger than the mainshock, the aftershock is redesignated as the mainshock and the original main shock is redesignated as a foreshock. Aftershocks are formed as the crust around the displaced fault plane adjusts to the effects of the mainshock.

Earthquake swarms

Earthquake swarms are sequences of earthquakes striking in a specific area within a short period. They are different from earthquakes followed by a series of aftershocks by the fact that no single earthquake in the sequence is obviously the main shock, so none has a notable higher magnitude than another. An example of an earthquake swarm is the 2004 activity at Yellowstone National Park.

In August 2012, a swarm of earthquakes shook Southern California's Imperial Valley, showing the most recorded activity in the area since the 1970s.

Sometimes a series of earthquakes occur in what has been called an earthquake storm, where the earthquakes strike a fault in clusters, each triggered by the shaking or stress redistribution of the previous earthquakes. Similar to aftershocks but on adjacent segments of fault, these storms occur over the course of years, and with some of the later earthquakes as damaging as the early ones. Such a pattern was observed in the sequence of about a dozen earthquakes that struck the North Anatolian Fault in Turkey in the 20th century and has been inferred for older anomalous clusters of large earthquakes in the Middle East.

Intensity of earth quaking and magnitude of earthquakes

Quaking or shaking of the earth is a common phenomenon undoubtedly known to humans from the earliest times. Before the development of strong-motion accelerometers that can measure peak ground speed and acceleration directly, the intensity of the earth-shaking was estimated based on the observed effects, as categorized on various seismic intensity scales. Only in the last century has the source of such shaking been identified as ruptures in the Earth's crust, with the intensity of shaking at any locality dependent not only on the local ground conditions but also on the strength or magnitude of the rupture, and on its distance.

The first scale for measuring earthquake magnitudes was developed by Charles F. Richter in 1935. Subsequent scales (see seismic magnitude scales) have retained a key feature, where each unit represents a ten-fold difference in the amplitude of the ground shaking and a 32-fold difference in energy. Subsequent scales are also adjusted to have approximately the same numeric value within the limits of the scale.

Although the mass media commonly reports earthquake magnitudes as "Richter magnitude" or "Richter scale", standard practice by most seismological authorities is to express an earthquake's strength on the moment magnitude scale, which is based on the actual energy released by an earthquake.

Frequency of occurrence

It is estimated that around 500,000 earthquakes occur each year, detectable with current instrumentation. About 100,000 of these can be felt.

Minor earthquakes occur nearly constantly around the world in places like California and Alaska in the U.S., as well as in El Salvador, Mexico, Guatemala, Chile, Peru, Indonesia, the Philippines, Iran, Pakistan, the Azores in Portugal, Turkey, New Zealand, Greece, Italy, India, Nepal and Japan.

Larger earthquakes occur less frequently, the relationship being exponential; for example, roughly ten times as many earthquakes larger than magnitude 4 occur in a particular time period than earthquakes larger than magnitude 5.

In the (low seismicity) United Kingdom, for example, it has been calculated that the average recurrences are: an earthquake of 3.7–4.6 every year, an earthquake of 4.7–5.5 every 10 years, and an earthquake of 5.6 or larger every 100 years.

This is an example of the Gutenberg–Richter law.

The Messina earthquake and tsunami took as many as 200,000 lives on December 28, 1908, in Sicily and Calabria.

The number of seismic stations has increased from about 350 in 1931 to many thousands today. As a result, many more earthquakes are reported than in the past, but this is because of the vast improvement in instrumentation, rather than an increase in the number of earthquakes. The United States Geological Survey estimates that, since 1900, there have been an average of 18 major earthquakes (magnitude 7.0–7.9) and one great earthquake (magnitude 8.0 or greater) per year, and that this average has been relatively stable.

In recent years, the number of major earthquakes per year has decreased, though this is probably a statistical fluctuation rather than a systematic trend. More detailed statistics on the size and frequency of earthquakes is available from the United States Geological Survey (USGS). A recent increase in the number of major earthquakes has been noted, which could be explained by a cyclical pattern of periods of intense tectonic activity, interspersed with longer periods of low intensity. However, accurate recordings of earthquakes only began in the early 1900s, so it is too early to categorically state that this is the case.

Most of the world's earthquakes (90%, and 81% of the largest) take place in the 40,000-kilometre-long (25,000 mi), horseshoe-shaped zone called the circum-Pacific seismic belt, known as the Pacific Ring of Fire, which for the most part bounds the Pacific Plate. Massive earthquakes tend to occur along other plate boundaries too, such as along the Himalayan Mountains.

With the rapid growth of mega-cities such as Mexico City, Tokyo and Tehran in areas of high seismic risk, some seismologists are warning that a single quake may claim the lives of up to three million people.

Induced seismicity

While most earthquakes are caused by movement of the Earth's tectonic plates, human activity can also produce earthquakes. Activities both above ground and below may change the stresses and strains on the crust, including building reservoirs, extracting resources such as coal or oil, and injecting fluids underground for waste disposal or fracking. Most of these earthquakes have small magnitudes. The 5.7 magnitude 2011 Oklahoma earthquake is thought to have been caused by disposing wastewater from oil production into injection wells, and studies point to the state's oil industry as the cause of other earthquakes in the past century. A Columbia University paper suggested that the 8.0 magnitude 2008 Sichuan earthquake was induced by loading from the Zipingpu Dam, though the link has not been conclusively proved.

Measuring and locating earthquakes

The instrumental scales used to describe the size of an earthquake began with the Richter magnitude scale in the 1930s. It is a relatively simple measurement of an event's amplitude, and its use has become minimal in the 21st century. Seismic waves travel through the Earth's interior and can be recorded by seismometers at great distances. The surface wave magnitude was developed in the 1950s as a means to measure remote earthquakes and to improve the accuracy for larger events. The moment magnitude scale not only measures the amplitude of the shock but also takes into account the seismic moment (total rupture area, average slip of the fault, and rigidity of the rock). The Japan Meteorological Agency seismic intensity scale, the Medvedev–Sponheuer–Karnik scale, and the Mercalli intensity scale are based on the observed effects and are related to the intensity of shaking.

Every tremor produces different types of seismic waves, which travel through rock with different velocities:

Longitudinal P-waves (shock- or pressure waves)

Transverse S-waves (both body waves)

Propagation velocity of the seismic waves through solid rock ranges from approx. 3 km/s (1.9 mi/s) up to 13 km/s (8.1 mi/s), depending on the density and elasticity of the medium. In the Earth's interior, the shock- or P-waves travel much faster than the S-waves (approx. relation 1.7:1). The differences in travel time from the epicenter to the observatory are a measure of the distance and can be used to image both sources of quakes and structures within the Earth. Also, the depth of the hypocenter can be computed roughly.

In the upper crust, P-waves travel in the range 2–3 km (1.2–1.9 mi) per second (or lower) in soils and unconsolidated sediments, increasing to 3–6 km (1.9–3.7 mi) per second in solid rock. In the lower crust, they travel at about 6–7 km (3.7–4.3 mi) per second; the velocity increases within the deep mantle to about 13 km (8.1 mi) per second. The velocity of S-waves ranges from 2–3 km (1.2–1.9 mi) per second in light sediments and 4–5 km (2.5–3.1 mi) per second in the Earth's crust up to 7 km (4.3 mi) per second in the deep mantle. As a consequence, the first waves of a distant earthquake arrive at an observatory via the Earth's mantle.

On average, the kilometer distance to the earthquake is the number of seconds between the P- and S-wave times 8. Slight deviations are caused by inhomogeneities of subsurface structure. By such analysis of seismograms, the Earth's core was located in 1913 by Beno Gutenberg.

S-waves and later arriving surface waves do most of the damage compared to P-waves. P-waves squeeze and expand the material in the same direction they are traveling, whereas S-waves shake the ground up and down and back and forth.

Earthquakes are not only categorized by their magnitude but also by the place where they occur. The world is divided into 754 Flinn–Engdahl regions (F-E regions), which are based on political and geographical boundaries as well as seismic activity. More active zones are divided into smaller F-E regions whereas less active zones belong to larger F-E regions.

Standard reporting of earthquakes includes its magnitude, date and time of occurrence, geographic coordinates of its epicenter, depth of the epicenter, geographical region, distances to population centers, location uncertainty, several parameters that are included in USGS earthquake reports (number of stations reporting, number of observations, etc.), and a unique event ID.

Although relatively slow seismic waves have traditionally been used to detect earthquakes, scientists realized in 2016 that gravitational measurements could provide instantaneous detection of earthquakes, and confirmed this by analyzing gravitational records associated with the 2011 Tohoku-Oki ("Fukushima") earthquake.

Bibliography

- D., Talley, Lynne; L., Pickard, George; J., Emery, William; (Oceanographer), Swift, James H. (2011). Descriptive physical oceanography : an introduction. ISBN 9780750645522. OCLC 784140610.
- Pinet, Paul R. (1996). Invitation to Oceanography (3rd ed.). St. Paul, MN: West Publishing Co. ISBN 0-7637-2136-0.
- Hamblin, W. Kenneth; Christiansen, Eric H. (1998). Earth's Dynamic Systems (8th ed.). Upper Saddle River: Prentice-Hall. ISBN 0-13-018371-7.
- Marshak, Stephen (2001). Earth: Portrait of a Planet. New York: W.W. Norton & Company. ISBN 0-393-97423-5.
- Munk, W. and Wunsch, C., 1998: Abyssal recipes II: energetics of tidal and wind mixing. Deep-Sea Research Part I, 45, pp. 1977--2010.
- Talley, Lynne D. (Fall 2013). Reading-Advection, transports, budgets. SIO 210: Introduction to Physical Oceanography. San Diego: Scripps Institute of Oceanography. University of California San Diego. Retrieved August 30, 2014.
- Macdonald, Alison M. (1995). Oceanic fluxes of mass, heat, and freshwater : a global estimate and perspective. WHOI Theses. Falmouth, Mass.: Massachusetts Institute of Technology and the Woods Hole Oceanographic Institution. p. 12. hdl:1912/5620.
- Su, Jingzhi; Li, Tim; et al. (2014). The Initiation and Developing Mechanisms of Central Pacific El Niños. Journal of Climate. 27 (12): 4473-4485. doi:10.1175/ JCLI-D-13-00640.1.
- Hansen, B.; Østerhus, S; Quadfasel, D; Turrell, W (2004). Already the day after tomorrow?. Science. 305 (5686): 953-954. doi:10.1126/science.1100085. PMID 15310882. S2CID 12968045.
- Kerr, Richard A. (2004). A slowing cog in the North Atlantic ocean's climate machine. Science. 304 (5669): 371-372. doi:10.1126/science.304.5669.371a. PMID 15087513. S2CID 42150417.
- Lahr, J. C.; Chouet, B. A.; Stephens, C. D.; Power, J. A.; Page, R. A. (1994). Earthquake classification, location, and error analysis in a volcanic environment: implications for the magmatic system of the 1989-1990 eruptions at Redoubt Volcano, Alaska (PDF). Journal of Volcanology and Geothermal Research. 62 (1-4): 137-151. Bibcode:1994JVGR...62..137L. doi:10.1016/0377-0273(94)90031-0.

- White, R; McCausland, W (2016). Volcano-tectonic earthquakes: A new tool for estimating intrusive volumes and forecasting eruptions. Journal of Volcanology and Geothermal Research. 309: 139-155. Bibcode:2016JVGR..309..139W. doi:10.1016/j.jvolgeores.2015.10.020. ISSN 0377-0273.
- Roman, D. C.; Cashman, K. V. (2006). The origin of volcano-tectonic earthquake swarms. Geology. 34 (6): 457-460. Bibcode:2006Geo....34..457R. doi:10.1130/G22269.1.
- Schmincke, H. U. (2004). Volcanism. doi:10.1007/978-3-642-18952-4. ISBN 978-3-642-62376-9.
- McNutt, S. R. (1996), Seismic Monitoring and Eruption Forecasting of Volcanoes: A Review of the State-of-the-Art and Case Histories, in Scarpa, Roberto; Tilling, Robert I. (eds.), Monitoring and Mitigation of Volcano Hazards, Springer Berlin Heidelberg, pp. 99-146, doi:10.1007/978-3-642-80087-0_3, ISBN 9783642800870
- Guerriero, V; Mazzoli, S. (2021). Theory of Effective Stress in Soil and Rock and Implications for Fracturing Processes: A Review. Geosciences. 11 (3): 119. Bibcode:2021Geosc..11..119G. doi:10.3390/geosciences11030119.
- Nur, A; Booker, J.R. (1972). Aftershocks Caused by Pore Fluid Flow?. Science. 175 (4024): 885-887. Bibcode:1972Sci...175..885N. doi:10.1126/science.175.4024.885. PMID 17781062. S2CID 19354081.
- What are Aftershocks, Foreshocks, and Earthquake Clusters?. Archived from the original on 2009-05-11.
- Earthquake Swarms at Yellowstone. United States Geological Survey. Retrieved 2008-09-15.
- Duke, Alan. Quake 'swarm' shakes Southern California. CNN. Retrieved 27 August 2012.
- Amos Nur; Cline, Eric H. (2000). Poseidon's Horses: Plate Tectonics and Earthquake Storms in the Late Bronze Age Aegean and Eastern Mediterranean (PDF). Journal of Archaeological Science. 27 (1): 43-63. doi:10.1006/jasc.1999.0431. ISSN 0305-4403. Archived from the original (PDF) on 2009-03-25.
- Earthquake Storms. Horizon. 1 April 2003. Retrieved 2007-05-02.
- Goldman, Jana (March 20, 2012). Amount of coldest Antarctic water near ocean floor decreasing for decades. NOAA. Retrieved 30 August 2014.
- Physical Oceanography Archived 2012-07-17 at archive.today Oregon State University.
- MyWorldCat list-OceanHeat. WorldCat. Retrieved Aug 30, 2014.
- Aftershock | geology. Encyclopedia Britannica. Retrieved 2021-10-13.
- Repeating Earthquakes. United States Geological Survey. January 29, 2009. Retrieved May 11, 2009.

Index

J

L

M

N

O

P

Q

R

S